W9-CCC-678

Galileo Galilei – When the World Stood Still

Atle Næss

Galileo Galilei – When the World Stood Still

With 21 Illustrations

Atle Næss
e-mail: alte.naess@sensewave.com

Translator
James Anderson
e-mail: JSINCLAIRA@aol.com

Library of Congress Control Number: 2004114622

The cover picture is printed with permission from akg images.

Author photo by Astrid M. Ledang.

This translation has been published with the financial support of NORLA Non-fiction.

Translation of the 2nd edition of "Da Jorden stod Stille" by Atle Næss, © Gyldendal Norsk Forlag AS, Oslo, 2002.

ISBN 3-540-21961-7 Springer Berlin Heidelberg New York

Springer is a part of Springer Science+Business Media

springeronline.com

Printed in Germany

Production and typsetting: LE-TEX Jelonek, Schmidt & Vöckler GbR, Leipzig
Cover design: *design & production* GmbH, Heidelberg

Printed on acid-free paper SPIN 11007227 46/3142YL - 5 4 3 2 1 0

This book about a man who rejected ancient truths in his quest for new knowledge is dedicated to the two most inquisitive people I know: my daughters Åshild and Unn Gyda.

Contents

Prologue: A Journey to Rome

When the greatest scourge of the Catholics, Gustav Adolf, the 'Lion of the North', fell at the battle of Lützen in the autumn of 1632, that grim war north of the Alps had raged for fourteen years. All across Catholic Europe thanksgiving masses were said. When news of the Swedish King's death reached Rome, His Holiness Pope Urban VIII ordered a Te Deum to be performed in the Sistine Chapel, and he himself sang the versicles.

Most of the inhabitants of the Italian states also gave thanks to God, glad to have avoided the war itself and the great, destructive bands of soldiers that plundered and starved whole regions. But this did not mean the Italians had been spared misfortunes of every sort. War's sinister step-brother, the plague, was ravaging the peninsula.

The Grand Duchy of Tuscany and its capital, Florence, were severely affected. Everyone knew the symptoms: sufferers were stricken with faintness, and after a few hours black buboes appeared in the groin and armpits. The buboes were a sure sign. Everyone knew then what to expect. The sick and their relatives could do little more than wait. And they did not have to wait long. Soon, dark spots appeared all over the body, followed by a high fever, the bouts of bloody vomiting and a swift, certain death.

In the small village of Arcetri, on a wooded hillside just south of Florence, an old man sat writing his will. He had to make a journey to Rome and wanted to be prepared for every eventuality. If the plague did not get him on the road, the strain of travelling might finish him off; in addition he had been ill most of the autumn, with dizziness, stomach pains and a serious hernia. And even if he survived these difficulties, and the cold winter wind from the Apennines did not give him pneumonia, he had no idea what awaited him in Rome, only that his arrival was unlikely to be celebrated with a special mass.

He had attempted to put off the journey all the previous autumn by pleading that he was elderly and frail. It had made not the least difference; if anything it had irritated his powerful enemies even more. The last summons he had received had been quite unambiguous: if he did not come instantly of his own volition he would be arrested, put in chains and taken away despite his advanced age and high standing.

He walked the short distance through the bare cornfields and vineyards to visit his two daughters. Both were nuns at the convent of San Matteo, married only to Christ. He had personally been instrumental in this. Only a couple of years ago he had moved to the villa in Arcetri, to be closer to both of them. Now he was not sure if he would ever meet them again. But he knew they would pray for him, and that their prayers might be needed.

Next, he sent a summons to his only son and his two small grandchildren, both boys, so that he could take his leave of them. The elder of the boys had just turned three and had been christened after him. The will that he had just made named his son as his sole heir.

The old man's employer and protector was the youthful Grand Duke of Tuscany. Although the name of Medici still commanded some respect, the 22-year old ruler could do nothing to prevent his ageing mathematician and philosopher from having to make this humiliating and dangerous journey. But the Grand Duke provided the most comfortable means of travel at his disposal, a commodious carriage from the grand ducal carriage houses. The trip would still take at least a fortnight, but it would ease the strain on the old man a little.

On 20 January 1633, he set out southwards from Florence. After a couple of days' travelling through the Chianti region he arrived at Siena, where he had spent a winter during his youth, almost half a century earlier. Now wind and sleet blew across the brick-red, amphitheatre-like city square, and he had no time to relive old memories. He continued slowly southwards through the great chestnut forests on the slopes of Monte Amiata, the mountain that forms an almost perfect cone as it rises steeply above the low wooded hillsides that surround it.

When he got to Ponte a Centina near the little border town of Acquapendente, a nasty surprise greeted him. Because of the plague no one was allowed into the Papal States without fourteen days' quarantine. Sleeping accommodation was pitiful and it was hard to buy food. He managed to get bread and wine, and occasionally a few eggs. His orders had been to come to Rome as quickly as possible, and the old man believed he had been given exemption from quarantine. But the border guards had their orders: no exceptions regardless of errand.

Finally he was able to proceed, past Lake Bolsena, down to Viterbo and on to the Via Cassia, one of the many roads that radiated from the ancient city of Rome. That soon took him into the city.

He arrived in Rome on 13 February. It was the first Sunday in Lent and two days before his sixty-ninth birthday. Here, one small consolation awaited him: he was to be the guest of the Grand Duke's Ambassador until his case came up.

The impressive villa on the slopes of Monte Pincio conjured up memories of happier visits to Rome, when his name had been on the lips of everyone in the city and all of them – professors, cardinals, noblemen, even His Holiness himself – wanted to hear about his theories and discoveries. Now the Embassy had become a benign prison. But at least for the time being he was spared real imprisonment. This gave him the slender hope that everything might yet be sorted out amicably.

Hope grew as week after week went by and the Ambassador appeared to work assiduously on his behalf. The spring came, he could sit in the great park that surrounded the villa, and enjoy elevated views right across the city to St. Peters on the far side of the Tiber and admire the dome that his great Tuscan compatriot Michelangelo had constructed. But he was racked with rheumatism, and the news from his family back home in Florence was troubling: the plague had flared up once more. The Florentines heard the constant ringing of small bells in the evening darkness, announcing that the corpse-bearers were at work.

In reality, the Ambassador achieved little by his enquiries, other than to gain time. But he revealed none of this in order to spare the old man as much anxiety as possible. Finally, on 9 April, the summons came: the Grand Duke's mathematician and philosopher, signor Galileo Galilei, had to appear before the Holy Office, also known as the Inquisition, in three days' time. There he would be interrogated and incarcerated for an indefinite period, until judgement was delivered in the case against him.

The Musician's Son

The detached belfry of Pisa Cathedral leant dangerously southwards. It looked peculiar, but the phenomenon attracted no attention outside the city itself. Tuscans were used to ostentatious towers on both private and public buildings, and it was accepted that, from time to time, one or other of them might come crashing to the ground.

This zealous tower building encapsulated two of the traits characteristic of the Tuscan: firstly, his intense need to draw attention to himself, quite literally to raise himself above others. Secondly, his almost miraculous combination of craftsmanship, technical expertise and artistic talent which had made Tuscany, and particularly its capital Florence, into the Western World's undisputed centre for architecture, sculpture and painting during an age that an admiring future was to christen the Renaissance.

This golden age was definitely on the wane by the year 1564.

Cosimo I de' Medici was Duke of Tuscany. The Medicis had originally been physicians, but had later turned to banking and business. For more than a century the family had dominated Florence with its power and wealth. But new times had arrived in Europe, an age of absolute monarchy, and power had to be legitimised by reference to a ruler's noble lineage and divine right. Cosimo had acquired a ducal title and established himself as absolute ruler. He had moved from the Palazzo Vecchio in the city's ancient, pulsating centre, across the river Arno to the huge and enclosed Palazzo Pitti. There, at a regal distance from the humdrum life of the city, the Duke and his court lived with a pomp that would have been the envy of many a European king.

The musician, Vincenzio Galilei, was the same age as Cosimo de' Medici. He too came from an old Florentine family with a medical ancestor. There, any similarity with the Medicis abruptly ceased. Wealth and power had notably eluded the Galilei family.

The Duke's court was a place of work for Vincenzio, an arena in which he could play the lute and viola da gamba. But he could not get enough commissions there or in Florence as a whole. Things got even more difficult when he married Giulia, a woman twenty years his junior. Her family came from Pisa, and Vincenzio felt forced to move there. This was no easy decision for a patriotic Florentine. But the cost of living was lower in Pisa, a musician had less competition there and, above all, his wife had family in the city, practical, hard-working folk in the woollen trade who could offer a poor relation a little work now and again.

The bond between Florence and Pisa had never been very cordial. In his *Divine Comedy*, Florence's greatest son, Dante Alighieri, depicts Pisa as the cradle of treachery, and places some very eminent Pisans in the deepest depths of Hell. But the two cities were no longer rivals of equal rank. From its position as one of Europe's richest and most powerful city states, Pisa had degenerated into a sleepy Tuscan provincial town, firmly ruled from Florence.

Vincenzio had married to keep the Galilei family going: his Giulia was pregnant. On 15 February 1564 the couple's eldest son was born in a rented house near the church of Sant' Andrea, half way between the university and the Medicis' local palace. Following a relatively common Tuscan tradition, the boy was given the singular form of the family name as a Christian name: Galileo. He was called after the original 15th century founder of the line, the doctor now buried in no less a place than the church of Santa Croce.

Vincenzio Galilei was not only a skilled musician and noted composer. He was a learned man. What interested him most was the theory of music. He had studied with well known humanists in Venice and Rome, and was engaged in writing a great thesis in which he was ambitiously attempting to revive contemporary music by returning to the principles of antiquity.

Young Galileo was not an only child. His mother Giulia gave birth to six more children in rapid succession, but only one brother and two sisters lived to adulthood. Vincenzio soon realised that his eldest son was uncommonly gifted and lavished special attention on him. He taught Galileo to play the lute, and the boy soon became a skilful player.

He also learnt two other things from his father's toil with his thesis. The first was that one should never be content with accepted wisdom, even if it came from the most authoritative sources, but combine theoretical deliberations with practical experiments and arrive at one's own conclusions.

The second was that such pioneering work was often, quite literally, undervalued. Vincenzio constantly struggled to provide for himself and his family. In 1572 he moved back to Florence alone. Cosimo had just been elevated to *Grand* Duke, and the celebrations offered an opportunity for a good musician to shine at court. But Giulia and the children had to remain with her family in Pisa, and it is tempting to imagine young Galileo overhearing his mother's relatives making remarks about who had to support him and his brother and sisters.

In 1574 Grand Duke Cosimo died. He was a temperamental tyrant who once killed a servant on the spot because he had told Cosimo's son that his father was considering re-marrying; but he was also a generous patron and enterprising ruler who had brought material prosperity to his central Italian Grand Duchy. The majority of Tuscans harboured no high expectations of his son, Francesco. Their worst fears were realised. Francesco's spouse died under mysterious circumstances, after which he held an extravagant wedding ceremony with his infamous lover, Bianca. Even worse was the fact that the new Grand Duke protected his younger brother Pietro, who had strangled his wife in a fit of jealousy.

It was at this court that Vincenzio was to earn most of his living. The change of grand dukes did not alarm him, for he brought Giulia and his children to live with him in Florence. The family settled close to one of the bridges over the Arno, Ponte delle Grazie. It was a practical place to live. The Grand Duke's Palazzo Pitti lay close by.

Ten-year old Galileo had come home. His family belonged in Florence. Ever after he considered himself to be a Florentine. But his father was not satisfied with the education the boy could receive in the city of his ancestors. The following year he sent Galileo to the remote monastery at Vallombrosa – the "shady valley" – north of Regello in Valdarno, some twenty miles southeast of Florence.

The contrast with a city like Florence could hardly have been greater. The monastery was beautifully situated, but was completely isolated and at an elevation of over 3,000 feet, surrounded by a forest of broad-leaved trees as well as heavy, dark spruces with ivy-clad trunks.

Vincenzio knew what he was doing. The monks of this monastery belonged to the intellectual Florentine tradition. It was an inspiring environment, far beyond the standard of monasteries generally. Here, the gifted young boy could learn Greek, Latin and logic.

Galileo was an assiduous student who thoroughly enjoyed life in these isolated, spartan surroundings. But the boy liked it even better than his father had hoped. After a couple of years he wanted to join the order, and offered himself as a novice.

Perhaps it was youthful religious passion that lay behind this decision, but Galileo also perceived that the strict life of a monk would provide him with opportunities for work and study, free from the material cares that the life of a citizen brought with it. Vincenzio, however, had no sympathy with his eldest son's decision. In 1579, he took the winding mountain road up to the monastery and brought the fifteen-year-old back home to Florence.

His father's motives may have been to prevent Galileo becoming stuck in a location and environment which, in the long run, would never be able to provide him with sufficient challenges. But it is more likely that cold financial calculations lay behind this "rescue expedition". Vincenzio would have to make contributions to the running costs of the monastery if his son were to become a monk. *Daughters* might feasibly be candidates for monastic life. They had to be subsidised as well, of course, but if they married instead, their father had to find a dowry, so daughters were costly in any event. But a son like Galileo ought to find himself paid work, so that he could help out with the family's expenses.

But what career was his son to choose?

A Gifted Young Tuscan

Galileo Galilei was an impoverished young man with big ambitions and many talents. He was to prove a brilliant writer. He was musical like his father. He could draw and paint, and he seriously considered making his livelihood in art – a career that traditionally was very prestigious in Florence, where training opportunities were second to none.

Galileo well knew what an artist's life was like. It was at about this time that he struck up a close friendship with Lodovico Cardi, known by the name Cigoli, who was barely five years his senior. At an early age this gifted painter was commissioned by the Medici family and was rated as the finest among his contemporaries in Florence.

His father's work and his own environment inclined Galileo more towards art than to science. But in the wake of the Renaissance, the line between these two areas was not very clearly defined. Vincenzio's musical theory made use of mathematics and physics – indeed, music as a taught subject was reckoned as one of the quadrivium subjects, together with arithmetic, geometry and astronomy. (The linguistic disciplines – trivium – were grammar, rhetoric and logic.) Painting was seen as closely related to geometry, principally because of the theory of perspective. It was taken as read that painters had to study anatomy. The young Cigoli was so keen on dissection that he contracted a serious and long-lasting illness through over-exposure to cadavers!

Vincenzio, however, was not enthusiastic about his son's artistic pretensions. He knew only too well what kind of existence such a life had to offer. And painting was at least as insecure as music. His father had a better idea. Galileo was to study medicine and become a prosperous doctor, like their ancestor. Good son that he was, Galileo laid his painting ambitions aside and obeyed his father's wishes.

Medicine was far from being a poor career choice for a young man with ambitions. The discipline was particularly prestigious in Italy, whereas in most other European countries theology still dominated the universities. It was a comprehensive education. In those days subject boundaries were not clear cut – it is questionable if "disciplines" in the modern sense existed at all. Natural philosophy, logic and mathematics were "medical subjects", as well as the very recently developed anatomy, with its spectacular dissections. Mathematics and astronomy were important for doctors principally because they had to be able to cast accurate horoscopes for their patients. They had little more in their armoury with which to fight serious disease.

Galileo returned to his native city, Pisa, in 1581 as a 17-year old student. He had come to the provinces. The city's hub, Piazza dei Cavalieri, could not compare either in size or liveliness with the Piazza Signoria in Florence, even though its beautiful palace boasted fine external frescos by Cosimo's court painter, Vasari . Similarly, the intellectual life of the University of Pisa was nothing like that of centres like Bologna or Padua. It was an educational establishment without international cachet, where the average professor was as interested in his social status as in academic achievement.

Galileo began to attend the lectures that were relevant to medicine, and it was not long before it became apparent that he was no ordinary student. He was not content to repeat his teachers' dogmatic interpretation of accepted truths.

It is said that Galileo's first scientific discovery was made in Pisa Cathedral during Mass. From his pew in the church he noticed a chandelier that was swinging to and fro, and he noted that the time these small oscillations took was constant and unrelated to how far the lamp swung.

This observation would, many years later, lead to the construction of the pendulum clock and a hitherto unknown accuracy in the measurement of time. But in the first instance the young medical student and some friends made a simpler contrivance, a so-called *pulsilogium*. The measurement of pulse was an important diagnostic tool for the doctors of that period. Galileo constructed a pendulum, the length of which could be adjusted so that it swung in time with the patient's pulse. Now the doctor could read a diagnosis directly from the length of the pendulum!

*

In 1583 Grand Duke Francesco came to Pisa as usual, where his court spent their time between Christmas and Easter. The Medici family had owned a palace there for many years, and Francesco began the building of a newer and larger one, in the best district, down by the Arno. In this way he could add lustre to the city and remind the Pisans of who held power in Tuscany.

Grand Duke Francesco's retinue contained a mathematician and military engineer by the name of Ostilio Ricci. He came into contact with Galileo and discovered that the young student was interested in mathematics.

The teaching of mathematics at the university was extremely poor. The subject had a low status compared to general natural philosophy. Ricci opened a new world to the young student, the world of algebra and geometry. He made Galileo acquainted with the works of a Venetian named Niccolò Tartaglia, who had probably been Ricci's own teacher, and who was regarded as the greatest Italian mathematician of the 16th century.

Tartaglia left his mark on the history of mathematics. He was the first to find a general method of solving cubic equations. Galileo, however, skimmed rather quickly through this new arithmetic, even though it clearly had practical applications. He did precisely as his father had done in the musical sphere, he turned to the inheritance from antiquity. As far as mathematics was concerned this meant the rediscovery of Euclid and Archimedes . It was this traditional, classical mathematics with its strong emphasis on geometry, that fascinated him. And it was Ricci who opened his eyes to this aspect of Tartaglia's work as well: Tartaglia had in fact translated, annotated and published Euclid and Archimedes in new editions and had thus made them accessible.

Galileo was a impecunious student, who sorely needed a lucrative profession. But the revelation that mathematics had opened up to him was more important then either his father's exhortations, or a possible future as a physician. It may also have helped that Ricci indicated a possible career path that would satisfy even the most ambitious: with the right contacts and the necessary skill one might end up as mathematician to a grand duke – a position that provided social rank and means beyond anything a doctor, or for that matter a professor, could aspire to.

Such an association with a court did of course also mean that any fall from grace would be a long one.

Vincenzio probably understood his son. He was working hard on his musical theory, and had finally completed his great thesis in dialogue form (*Dialogue on Ancient and Modern Music*). He argued polemically with his professional adversaries, while at the same time developing his theory in new directions with the aid of pure acoustic experiments.

But musical theory brought no money in. Vincenzio was simply unable to support his wife, three children and a student. In 1585 he had to ask Galileo to interrupt his studies at Pisa and return home to the Ponte delle Grazie, without a degree.

To Rome and the Jesuits

Galileo hurled himself into mathematics with an energy that showed he had finally found a calling, a direction to his life. Even without a degree he was undoubtedly one of the most knowledgeable men in Italy regarding mathematics. But this was of little use unless his talents were recognised. At home in Florence there was no mathematical set. He did a bit of private tutoring and spent one winter in Siena. In order to get on he had to make contacts.

With this in mind, Galileo set out on his first journey to Rome.

The Rome to which the young Florentine mathematician came in the autumn of 1587 was completely different to the Renaissance city where Rafael and Michelangelo had been great heroes earlier in the century. A lot had happened in the intervening period, the essence of which can be summed up in two words: Reformation and Counter-Reformation.

The papacy had strengthened its grip on the Church. Luther's Reformation in northern Europe was a seismic shockwave that demanded a new direction. The Council of Trent (1545–1563) spelt out the basic tenets of the Catholic faith, and at least got rid of some of the blemishes that Luther had pointed to. It was the start of a fight to win back its lost standing – the Counter-Reformation.

The Council of Trent accentuated the splits within Europe by defining the Catholic Church's ideological foundation: absolute monopoly on Christian teaching and interpretation. Every bit as important as the ideology was the inception of two executive organs to carry out the Counter-Reformation: the Jesuit Order (1540) and the reorganised ecclesiastical surveillance apparatus in the area of faith, the Roman Inquisition (1542). At the same time the popes began to view themselves more and more as absolute rulers; not merely as spiritual leaders, but also as princes of the Papal States, just like other sovereigns in autocratic Europe.

When Galileo arrived in Rome, he found himself in the midst of energetic upheaval in the city on various levels. Pope Sixtus V Peretti unrelentingly tore down cramped, old blocks of houses and constructed wide, straight thoroughfares between the main churches. The streets echoed to the constant noise of cobbles being pounded into place – more than a hundred streets were permanently surfaced in a five-year period.

And so Galileo could travel dry-shod over the cobblestones to the powerful, learned and influential organisation he had decided to contact – the Jesuits.

The young Jesuit Order had been founded in Paris by the Spanish nobleman, Ignatius Loyola. With a background in the army and higher education, Loyola built up within a few years an effective, elitist organisation that greatly emphasised teaching and scholarship, and which became the pope's strongest weapon against Luther's doctrines. Not least, the Jesuits achieved startling results in their missionary work, both in Asia and South America.

The two chief seats of the organisation's operations were in Rome and they had just been completed: the *Il Gesù* Church and the large, fortress-like centre of learning, *Collegio Romano*, which occupied an entire block in the middle of Rome between the Pantheon and the main street, Via del Corso.

In only a few years the Collegio Romano had become a very important institution and was considered to be one of the foremost universities of its age. When Galileo arrived there, 2,100 young men had either taken their degrees, or were still studying for them. There were also large Jesuit colleges in many other places including Köln, Trier and Munich.

Northern Europe was an important area of operations for the Jesuits, and there they undoubtedly helped to stem the tide of Lutheranism and Calvinism. The Jesuits literally conquered higher education. A key college was situated in Leuven (Louvain) in what is now Belgium, on the border between Catholic and Calvinist Europe. One of the Jesuit's keenest intellects, Robert Bellarmine, was at work there, but he would soon be returning to Rome to take up positions of even greater importance.

The Jesuits were famed for their somewhat unorthodox working methods, in which infiltration and undercover work was not unknown. One of Bellarmine's students at Leuven, a Norwegian called Laurits Nilssøn from Tønsberg, was sent to Protestant Stockholm, where – in the guise of a Protestant priest! – he built up an influential school and swayed King Johan III, who had married a Catholic, to such an extent that the King wanted to reintroduce Catholicism into the country, a notion that the clergy and his brothers soon put a stop to.

Galileo had not come to Rome and the college for religious reasons. The Jesuits had realised that if they wanted to wield influence, their scholastic calibre had to be of the very best, and the Collegio Romano could congratulate itself on possessing the greatest contemporary mathematician anywhere in Italy, Father Clavius.

Christopher Clavius was around fifty years of age. Originally German, he had been admitted to the Jesuit order at the age of seventeen and had spent most of his life in Italy. He wrote a number of textbooks on various mathematical and astronomical subjects, books that Galileo knew from his studies. He played a key part in the committee set up by Pope Gregory XIII which, just a few years before in 1582, had instigated a great reform. The result was the Gregorian Calendar, which is the foundation of our computation of time to this day. In brief, Father Clavius was a pivotal man to know for anyone wishing to make a career in mathematics on the mainland of Italy.

Totally unknown and unqualified, the 23-year old Tuscan was not overawed by the impressiveness of the Collegio Romano. He immediately sought out Father Clavius. Galileo explained his theories for calculating the centre of gravity of various objects, an area of study the Jesuit mathematicians were already interested in.

Clavius was impressed. He praised the practical work Galileo had done, and discussed the fundamental problems that arose as soon as mathematical models were transferred to the real, physical world: and indeed, whether this was even possible. The ideal, geometrical sphere touches a geometrical plane at just one point. But as soon as one uses a *real* sphere on a *real* plane, there is a contact *surface* of greater or lesser extent, between the two. As a result there were those who maintained that mathematics was, in a manner of speaking, self-absorbed; that it might indeed deliver incontrovertible proof, but only when dealing with abstracted mathematical subjects. Father Clavius, on the other hand, argued that mathematics was a necessary bridge between the abstract ("metaphysical") world and the one that actually existed.

Vincenzio Galilei's work on the relationship between string lengths and the perception of pitch reflected a practical attitude to mathematics as a working tool. Galileo's approach was the same, he showed this even as he watched his pendulum in Pisa Cathedral. This basic philosophy, that technical models could be used to reveal definite knowledge of the outside world, was strengthened by the ideas from the Collegio Romano. Presumably he was given lecture notes to take away with him and study at home in Florence.

His visit to Rome was proof of just how high Galileo was aiming. Working as a private tutor in his native city was to waste his time and talents. Nevertheless, Jesuit goodwill was not enough to secure him a permanent position. A professorship was vacant in Bologna, but it went to Giovanni Magini who was nine years older and had good connections with Duke Gonzaga in Mantua.

Galileo had to be content to travel back to Florence, to his family and his private lessons. But there were things happening in his native city: two sudden deaths. They set in motion a train of events that eventually would secure Galileo his first chair in mathematics.

A Surveyor of Inferno

It was rare for the citizens of Florence to see anything of their lord, Grand Duke Francesco de' Medici. He spent most of his time isolated in his villa in Pratolino with his extremely unpopular former mistress, now the Grand Duchess Bianca. Rumours in the city had it that they experimented with poisons which Bianca was to use in her murderous projects. The worst suspicions seemed to have been borne out when both of them died suddenly, on the same day in October 1587.

In fact, it was malaria that had killed them. At all events, that was the story of his brother and successor, and since Ferdinando was of a different stamp to Francesco, he was believed. Ferdinando de' Medici had been made a cardinal at the age of fifteen and had then spent many years in Rome, where he proved himself to be a womaniser of a somewhat unseemly sort for a churchman, but also a brilliant administrator and an avid collector of antique statues. He bought a large house on the slopes of Monte Pincio in order to have somewhere to store his collection. It was called the Villa Medici. But now he had to return home to Florence and his grand ducal title.

On the whole Ferdinando was a good ruler. He left the Church and married a distant relative. She was Christina of Lorraine, the granddaughter of King Henri II, a woman who was to be of great significance to Galileo. But more important for the mathematician's immediate future was Ferdinando's choice of his successor as cardinal.

It was generally accepted that a powerful family like the Medicis had to maintain their representation within the College of Cardinals. But now there was no suitable family member available. Instead, Grand Duke Ferdinando sought the election of a man he trusted – Francesco Maria del Monte.

The new Cardinal was not notably interested in questions of theology. Del Monte was a well educated aesthete, a man with a taste for the good life, but also seriously interested in poetry, art, music and science. He was well versed in Vincenzio Galilei's musical theory. Cardinal del Monte was not opulently rich, but lived very comfortably in the Palazzo Madama near the Piazza Navona. He liked latching on to promising young men and helping them – he was the first to discover Caravaggio's unruly artistic genius.

The Cardinal had a brother. His name was Guidobaldo and he was a mathematician.

During his visit to Rome, Galileo had become acquainted with Guidobaldo del Monte, although it did not help him very much in his quest for a position. Now, suddenly, the situation had drastically altered: Guidobaldo's brother was not only a cardinal, but was the Grand Duke's trusted man in Rome.

Galileo spoke to Guidobaldo, Guidobaldo to the Cardinal, the Cardinal to Grand Duke Ferdinando. The result was that in the autumn of 1589, Galileo could again return to his birthplace, Pisa, now as the 25-year old professor of mathematics.

But before leaving Florence, he gave a lecture in the city's prestigious Academy, founded to promote Tuscan as the foundation for the common Italian written language. He had been set the task of describing the location and dimensions of Dante's Hell. Florence was not a city to take its famous authors lightly. A well-known dramatist had once been exiled because he had announced that the sainted Catherine of Siena was a better writer than Florence's own Boccaccio!

The young freelance mathematician took his listeners by storm.

He was intimately versed in *The Divine Comedy* and the universe that was depicted there. Galileo explained the precise construction Dante had calculated for his Hell. It was shaped like a broad funnel, with its opening up on the surface of the earth. In each of its descending *circles* ever worse punishments were meted out to ever worse sinners, and using his skill in geometry, Galileo worked out the diameter of the various diabolical departments, in which various devils tortured the unhappy sinners for all eternity. The circles got narrower and narrower until they ended up at the centre of the Earth, where Lucifer himself reigned and everything was everlasting frost and ice – as far away from Heaven, light and warmth as it was possible to get.

Lucifer was at the centre of a sphere. Galileo did not need to produce arguments for this. His educated audience knew only too well that the earth was round. Every scholar had known that since antiquity. Eratosthenes of Alexandria had with fair accuracy calculated the circumference of the Earth 200 years before Christ – admittedly with a bit of luck in his assumptions. Thus Galileo had a starting point for estimating the relative dimensions.

In the matter of the *relationship* between the Earth and the rest of the universe, Dante, and all other learned men, held to a model that had been perfected by Ptolemy, another Greek from Alexandria, in the second century AD. Very briefly it can be described as follows: the Earth is the fixed and stable centre of the universe. Around it revolve the heavenly bodies at various distances, attached to invisible spherical shells – spheres – which propel them in circular orbits.

This *Ptolemaic* model seemed hardly more than plain and self-evident common sense – after all, that was how one experienced the Sun, Moon and stars. But Dante's universe was also a marvellous, ingenious alloy of cosmology and theology. Throughout the Middle Ages Ptolemy's thoughts had combined with theological ideas to form a mighty construction, in which God and his angels inhabited the different spheres – or heavens. The interplay between theology and astronomy was extremely intricate. For example, the tilt of the Earth's axis was explained by the Fall: as we know, this ended the state of paradise and brought transition and death into the world. God introduced the seasons and thus "the passage of time" by the simple expedient of tipping the Earth slightly out of its formerly "perfect" position.

But Galileo's subject was Hell. According to Dante, these funnel-shaped circles were created when Lucifer was thrown out of the upper reaches of Heaven, hit the Earth with great force – quite literally as a fallen angel – and then bored into the soil right to the centre of the sphere.

However, the young mathematician who had so impressed his fellow citizens with his understanding of Hell's dimensions, knew something that very few of his listeners had appreciated. An obscure canon by the name of Copernicus from the faraway Baltic coast, had developed a new theory. This theory was slowly permeating educated European circles. It was recklessly daring and could demolish the entire ingenious Ptolemaic edifice.

Galileo did not utter one word about this to the Academy in Florence, because something else was quite clear to him: such a huge cosmological and theological structure would never fall without resistance.

The Spheres from the Tower

The University of Pisa was situated close to the river Arno. The Medicis had built a fine rectangular building around an internal courtyard with a covered arcade, beneath which lecturers and students could stroll and argue in a dignified manner. The main subject for discussion, at least in the subjects concerned with natural philosophy, was Aristotle. His disciples had been called Peripatetics – those who walk about – because it was claimed that the master had taught in this way.

Aristotle's thoughts about the natural world had congealed into an unassailable system of instruction. In principle, his physics built on observation and the logical deductions arising from it. But the observations could be random and certainly were not systematised by means of controlled experiments. Emphasis was placed on the logical and philosophical conclusions – to such an extent that all the *practical* knowledge that had gradually accumulated, linked to technical advances in architecture and shipbuilding or the construction of clocks and the manufacture of spectacle lenses (to mention but a few), had barely impinged on university teaching of the fundamental physical questions concerning the natural world.

Many professors found greater academic prestige in interpreting an obscure passage of Aristotle than in observing for themselves. And academic discussion must adhere rigidly to the Master's pattern. It was still possible to hear, as a capping argument: *Ipse dixit!* – "He said so himself!" There were many, of course, who realised that not *every* answer to natural mysteries could be found in 1900-year old treatises, but nevertheless the Aristotelian framework of understanding limited their imagination and thought processes.

The very young Professor Galilei in occupying his chair at Pisa was not at all disconcerted that he had no degree himself. Thirty years later he was to write, comparing "good philosophers" to bad ones:

> "I believe (...) that they fly, and that they fly alone like eagles, and not like starlings [storni]. It is true that because eagles are scarce they are a little seen and less heard, whereas birds that fly in flocks fill the sky with shrieks and cries wherever they settle, and befoul the earth beneath them."[1]

No one should doubt that Galileo considered himself to be one of the eagles. While his older, Aristotelian colleagues flocked round their Master's books, the 25-year old sought new paths.

But he, too, found inspiration in a Greek thinker. Galileo's declared model was Archimedes. In addition, he was virtually an Italian, as he had lived and worked in Syracuse, a Greek colony in Sicily. Archimedes combined observation with rigorous deduction and achieved practical results from this. The famous law that bears his name was the result of a knotty problem he was set by the despotic ruler of Syracuse: to calculate the ratio of gold to silver in the king's crown.

By contrast with the logical and speculative Aristotle, Archimedes began harnessing the powerful tool of *mathematics* to calculate and describe physical processes. Galileo was professor of mathematics. He clearly saw that a fundamental uprating of the subject would give qualitatively better natural science.

The establishment at Pisa was interested in the principles of movement, that branch of physics which would later be called *kinematics*. One of his elder colleagues had written a huge work, *On Motion (De motu)* which was circulating in manuscript form. The author was quite clear that Aristotle's doctrine of motion was wanting in certain respects, but even so he could not manage to free himself from tradition.

The young, newly appointed Galileo was not especially impressed with *On Motion*. But instead of going on the offensive against this entire massive bastion of physical theory, he decided to aim at a single, but very moot point, one where observations could easily be made: he would describe a "heavy body" in "natural motion" – what we today would call "free fall".

Aristotle made two fundamental errors in his description of falling objects. Firstly, he maintained that any falling object would achieve a certain fixed speed, and secondly, that such speed was proportional to the weight of the object. Or, to put it another way: every falling object falls with a definite, "in-built" speed, the heavier the object the higher the speed.

Galileo demonstrated the absurdity of this last contention with a simple mental experiment. One takes two stones of similar weight and ties them together – now, all at once, they will fall twice as fast as they would separately! It is also flies in the face of all experience that a sphere weighing one kilo falls one metre in the same time it takes a ten kilo sphere to fall ten metres.

Galileo decide to investigate the matter from basic principles. Presumably he used – as his first biographer states – the obvious place for experiments in free fall: the famous detached, leaning belfry near the city's cathedral. In contrast to nearly everywhere else in Italy, the cathedral environs were not the city's main meeting place, but lay in peaceful seclusion by the north walls, so the chances of hitting passing townsfolk with falling iron balls was minimal.

He dropped wooden and iron balls, but the results of the experiments were far from conclusive. He could easily see that the balls fell at roughly the same speed, but that the iron ball hit the ground a little before the wooden one. He had no way of making precise observations, no clocks then were accurate enough to measure the fall times.

His observations were good enough to show that Aristotle's theories did not hold water, and Galileo tentatively launched his own. He concluded initially – and wrongly – that a body's falling speed is proportional to its mass density ("specific gravity"), a concept he had studied thoroughly in his work on Archimedes. He also realised that its speed was closely related to the medium it was falling through: an iron ball and a wooden ball might fall at roughly the same speed through air, but in water they behaved quite differently! Archimedes had taught him the concept of buoyancy, and this led him to reject yet another erroneous Aristotelian assumption: that bodies have an in-built "lightness" that operates in opposition to their "weight". The fact that wood floats in water is not due to its "lightness" lifting it up – it is simply that the material has a lower specific gravity than water.

However, for the time being he was saddled with the misconception that a falling body reaches a certain, stable speed of its own accord. It was then totally impossible, with the tools at his disposal, to measure the speed – far less the acceleration – of a sphere dropped from a tower.

Galileo did not only take Archimedes' point of view and argue for practical experiments to rebut Aristotle and inflexible academic thought, he also made sure he provoked his colleagues at Pisa on a more personal level.

Professorship brought with it the duty of donning a certain loose fitting official garb, based on the Roman toga. The young professor of mathematics had little time for the assumed and, to his mind, superficial dignity that this garment bestowed on its wearer. He penned a three-hundred-line lampoon[2] on the toga in all its essence. Not only could one trip up on such a garment, but as he pointed out, it also swathes the body in an impractical way. All clothing ought to be designed so that men and women could readily obtain an idea of each other's physical attributes, indeed: "it would be best to go about naked"! But worse still was the way the toga's dignity prevented the professors from visiting the brothel. That forced them, quite literally, to take the matter into their own hands – a pastime that was every bit as sinful as visiting a bordello, but considerably less satisfying.

And so Galileo made his mark as an oppositional paradoxer. It was impossible for him, as yet, to give written vent to this same colourful lack of respect in his own subject. Galileo actually wrote his own version of *On Motion*, but he did not try to get it printed. His free fall experiments were spectacular, but deficient, and it is probably a myth that the other professors and students gathered admiringly at the foot of the tower. There was still too much he did not understand.

From Pisa to Padua

Musician, composer and theoretician, Vincenzio Galilei, had married when he was more than forty years old. In 1591, that family-proud Florentine died at home in Florence. He had a permanent place in musical history, as well as a wife and four children, all of whom except Galileo and his sister Virginia, had no means of support.

The death meant that the young professor took over the responsibility for the entire family – a mother who was sometimes difficult and who was to live for another thirty years, a brother who was a minor and two sisters. His sister Virginia may have just got married, but a most important part of the marriage settlement had not been concluded: Vincenzio had not had the means to pay more than a fraction of the agreed dowry. The balance fell to Galileo – in regular instalments.

His younger sister, Livia, was just thirteen and was sent to a convent for the time being, but the convent cost money too. And his sixteen-year-old brother Michelangelo had, naturally, to continue the musical education he had begun.

As a newly appointed professor of mathematics Galileo earned 60 scudi per annum. It was almost a starvation wage. His colleagues in more prestigious fields were considerably better paid; professors of philosophy might earn up to 300–400 scudi. A really well-known painter could get 50 scudi for a single picture, or even 75 or one hundred in really favourable circumstances. A good doctor also brought in his 300 per annum.

These new responsibilities meant that he had to earn more money. The prospects for an imminent salary increase at Pisa were slender. Nor was the intellectual climate of his toga-clad colleagues especially inspiring with their stagnant Aristotelian dogmatics. Consequently, Galileo was most interested when a position at the University of Padua became vacant in the autumn of 1592.

Padua is not far inland from Venice, on the Po plain. The university was one of the oldest and most renowned in Italy, and was known as "Il Bo" – "The Bull", probably after an inn that reputedly stood close by. It was housed in an old palace and its banqueting hall was the scene of disputations and academic ceremonial. And, like Pisa, it had an internal quadrangle, which was surrounded by two storeys of colonnades above which the proud tower of the palace reared over staff and students alike.

From a scientific perspective the fact that Padua possessed Europe's oldest botanical gardens was of greater importance. Botany (like zoology) was a "progressive" science. The contact with America was a factor that contributed to the undermining of traditional natural history – it proved that there were many animal and plant species which neither Aristotle nor the other ancient authorities had known anything about. When Galileo arrived in the city, the botanical gardens at Padua had just taken delivery of an entirely new American species, which was being grown and observed with great interest. This was *soleanum tuberosum*, as it would later be called – also known as the potato.

The University of Padua was an intellectual powerhouse. This was partly because, as a seat of learning it had not been established by papal or imperial privilege, like most others. It had grown up out of the civic culture of the city and had what can only be called a "liberal profile". In 1564, Pope Pius IV had decreed that everyone who gained a degree from an Italian university, had to swear an oath of allegiance to Catholic doctrine. However, at Padua the university authorities managed to create loopholes in the provision that enabled northern European – Protestant – students to continue applying for places there.

It was at Padua that Vesalius had laid the foundations of modern anatomy with his controversial dissections, half a century before Galileo came to the city. During Galileo's time *The Bull* got its famous "anatomical theatre", complete with tribunes where students and other interested spectators could follow the dissections in detail. No less impressive is the fact that as early as 1678 Padua gave a degree to the world's first female university undergraduate, the philosopher Elena Lucrezia Cornaro.

Mathematics were another strong point. There was a number of applicants for the chair in mathematics, including the same Magini who a few years earlier had wrested Bologna from Galileo. Once again, Galileo had to count on his Roman contacts, the del Monte brothers. They originally came from Venice, and had influential friends both there and in Padua. In a concerted effort they managed to secure the post for Galileo – with a salary of 180 scudi, three times the rate at Pisa.

Padua belonged to the Venetian Republic. For centuries that powerful canal city had been laying claim to large areas of the hinterland. Galileo had to move away from his homeland in Tuscany, and as a servant of the state he required the permission of Grand Duke Ferdinando. This was graciously forthcoming.

In one sense Venice was quite similar to Florence: there, too, the golden age of architecture and art was drawing to a close. The city's greatest painter of all, Titian, was dead, after a career that spanned most of the 16th century. But Venice was still a republic, and its style was considerably more sober and civic than the Grand Duke's court. The authorities did not spend money on the ostentatious celebrations that Ferdinando in Florence had gradually become addicted to – preferably with a stage full of "volcanoes" and fire-spitting dragons. The Venetian Senate was more interested in sensible, public projects: the Rialto Bridge – as beautiful as it was practical – across the Canal Grande had just been completed in 1592.

Signs in the Sky

Venetian independence caused a slow, smouldering deterioration in the relationship with Rome and the ever more absolutist papal power. Neither the intellectuals nor the commonality of Venice were prepared to accept every decree from the papal throne uncritically. This was one of the reasons why the rootless, apostate Dominican friar Giordano Bruno chose to settle in Venice and Padua, when he made the foolhardy decision to return to Italian soil.

Giordano Bruno was a visionary and a philosopher, a charismatic thinker profoundly steeped in magic and ancient pantheism and, in the world of his own fantasy, well on the way to becoming a new Messiah. He was born in the little town of Nola near Naples and took to the life of a friar more for its educational possibilities than out of piety – it was philosophy that really interested him. In the 1570s he travelled to Rome, but had to flee the city on account of his many unorthodox views and what was no doubt a trumped-up charge of murder.

For more than fifteen years Bruno wandered about northern Europe, France, England and Germany. He gave lectures, disputed and wrote books. In Geneva he was arrested and expelled by the Calvinist authorities, in Toulouse he was allowed to teach at the university. King Henri III summoned him to Paris to learn about the extraordinary memorising techniques he had developed. Then he travelled to England, where he tried Oxford and later made contacts at court. He finally ended up in Germany – via France – where he went from university to university getting a reputation for being an all-knowing philosopher, but one without a firm religious commitment.

But then he wanted to go home. Giordano Bruno was a very talented mathematician, and he was in Padua to try to get the vacant professorship in mathematics.

Bruno wanted to substantiate his qualifications by giving private coaching in the city, but this strategy was unsuccessful. So by the time Galileo came to the city in the autumn of 1592, the friar had just left Padua, either because the chair had gone to a competitor, or for other reasons. After a couple of months in Venice, Bruno was denounced by his landlord, arrested and placed in the Inquisition's gaol.

The Roman Inquisition's stated objective was to fight all forms of heresy, and in principle, its jurisdiction covered the whole world. In practice, its power only really extended to the Italian states, where it functioned in tandem with the secular legal system. The inquisitors could themselves apprehend suspected persons, but more usually such people were turned over to them.

The system's pedantic efficiency – viewed in isolation – was legally unimpeachable. Its headquarters in Rome – *Sant'Ufficio*, the Holy Office – controlled its provincial courts, and ensured that practice was uniform everywhere, and there would be no hint of arbitrary justice with sentences being handed down according to the judges' whim and fancy. Painstaking minutes were kept, in which the notary was supposed to put down word for word everything that was said on both sides:

> "Not only all the defendant's responses and any statements he might make, but also what he might utter during the torture, even his sighs, his cries, his laments and tears."[3]

To begin with it looked as if Bruno would get over the problem by admitting to a few less important aberrations in matters of faith, and maintaining that, anyway, his stock-in-trade was philosophy and not religion. Despite centralisation, the local Inquisitor in Venice was not the worst person to deal with. But then the Inquisition's headquarters demanded that Bruno be sent to Rome. The secular Venetian authorities did precious little to prevent the extradition.

So began a process that was to last more than seven years. Bruno was thrown into the Holy Office's gaol not far from St Peter's. His literary works were many and not all were readily available, so the case rumbled slowly on, with interrogations and explanations. And so the situation remained, until the learned Jesuit, Cardinal Robert Bellarmine, took over the case. He cut through the chaff, specified eight heretical viewpoints that Bruno was purported to have promulgated in his writings, and asked him to repudiate them.

Bruno, isolated and by now confused, first agreed to this – then refused. The circumstances are unclear and the document has not survived, so we do not know exactly what the philosopher was sentenced for. It was said that he believed that Moses was a wizard, who fooled the Egyptians because he was more proficient than they were in the magic arts. Bruno also maintained that there must be an infinite number of universes, because anything else would be a limitation of God's omnipotence. This idea was viewed as heretical because it did not accord the Earth a central place in the universe.

Judgement was given on 8 February 1600. Giordano Bruno was sentenced as an "unrepentant heretic", "unyielding" and "obstinate". All of his works were placed on the register of prohibited books (*Index librorum prohibitorum*) as "heretical and erroneous and containing many heresies and false teachings".[4] Bruno was transferred to the "condemned cells" in the dungeon of Tor di Nona, on the east bank of the Tiber, directly opposite Castel sant'Angelo. He was taken from there on 17 February, after seven priests had seen him and tried to make him admit the error of his ways before the execution, which he refused to do. He was taken in an open cart, guarded by members of the Order of St John, who carried torches and intoned prayers.[5]

Giordano Bruno's last journey was made through the centre of Rome to the Flower Market, Campo de' Fiori, which was also the place of execution. Only the most important executions were carried out here, partly because the French Ambassador, who lived on the square, had complained about the sight and stench of the heretics' pyres.

But Bruno's execution *was* important. It was a reminder to everyone who had come to Rome for the holy jubilee, a reminder of the consequences of heresy. So the faggots stood waiting in the Flower Market, with bundles of twigs at the edge where the fire would be lit. The fifty-two-year-old Bruno was stripped naked and tied to the stake, the judgement was read to him and the outer, slender twigs were lit as prayers were said and psalms sung. A great crowd followed the awful progress of the fire as it licked up around the naked body.

Bruno's end in the Campo de' Fiori was by no means unique, his case is simply the most famous. The Inquisition did not distinguish between high and low, educated and uneducated. Just sixty or seventy miles north of Padua, for example, a case was proceeding just then against a humble miller who had had the misfortune to learn to read, and had formed a homespun concept of the world based on half understood fragments and his own perceptions. One of his many ideas was that the creation of the world was similar to the process of milk thickening into cheese. He too, ended up in the flames.

One of the many indictments that was raised during the process against Bruno, was that he believed that the Sun was static and that the Earth was a planet that moved through space just like the other planets. Giordano Bruno was, in other words, a Copernican.

De Revolutionibus Orbium Coelestium

In 1592, when Galileo arrived at Padua and Bruno was arrested, Mikolaj Copernik – or Copernicus – had been dead for almost fifty years, but the force of his ideas was only just beginning to make itself seriously felt.

In his private life Copernicus was hardly a revolutionary, he was in fact a peaceful cleric. He lived a quiet life as a canon of the cathedral in the small town of Frauenburg in the semi-independent bishopric of Ermeland on the shores of the Baltic, now part of Poland. As a young man around the year 1500 he had spent some years studying in Italy, thanks to a rich uncle. He had even been at *Il Bo* in Padua, although without attracting the least attention.

Copernicus took his doctorate in canon law. But he had studied many disciplines and his greatest interest lay in astronomy. He knew – as did all other learned men – that the accepted Ptolemaic view of the world was hard to reconcile with precise astronomical observations. In order to make the system function after a fashion, Ptolemy had to introduce a number of "auxiliary orbits" or *epicycles*, small circular orbits that the planets described on their journey around the Earth. On his deathbed in 1543, Canon Copernicus published a book – *De Revolutionibus Orbium Coelestium* – which tried to demonstrate that the description of the universe would be a lot simpler and more correct if one altered one basic concept: instead of assuming that the Sun, stars, planets and Moon travelled in circles and epicycles around a fixed Earth, one could conjecture that the *Earth* and the other planets orbited the Sun, while at the same time revolving on their own axes.

This idea was not quite original. It had been proposed by Greek philosophers, but Copernicus was the first to try to develop it systematically.

One might have assumed that Copernicus' revolutionary idea with the Sun in the centre (the *heliocentric* system) would have proved instantly compelling for all professional astronomers and that, at a stroke, everything would have fallen into place. But this was decidedly not the case. An example of the scepticism this theory aroused is shown in this muted British reaction to a lecture given by Giordano Bruno at Oxford:

> "Stripping up his sleeves like some jugler, and telling us much of *chentrum* & *chirculus* & *circumferenchia* (after the pronunciation of his country language) he undertooke among very many other matters to set on foote the opinion of Copernicus, that the earth did goe round, and the heavens did stand still; whereas in truth it was his owne head which rather did run round, & his braines did not stand still."[6]

The Copernican system was almost impossible to square with plain common sense. Anyone could raise objections to it: why do we not notice that the Earth is turning round, let alone hurtling through space at enormous speed? Not even educated men well versed in physics and astronomy had good answers. At the University of Copenhagen in Protestant Denmark, the astronomer Tycho Brahe had been one of the very first to lecture on Copernican theory in the winter of 1574–75. But Brahe himself was not convinced, and instead put together his own cosmological model.

One thing was perfectly clear to everyone who touched upon Copernicus' notion of moving the Earth from its place at the centre of the world and reducing it to one of several planets orbiting the Sun: they would encounter solid resistance from a united front of conservative natural philosophers and theologians. For one thing Ptolemy's system was considered part of Aristotle's description of physical reality, but a far worse problem was the Bible's own words. One needed to look no further than Holy Scripture's first page, Genesis, chapter 1, verses 17–18, which said unequivocally of the Sun, Moon and stars that: *God set them in the firmament of the heaven to give light upon the earth. And to rule over the day and over the night, and to divide the light from the darkness.* Not a word about the Earth making day and night by revolving.

Even so, the idea captivated some, both because of its remarkable simplicity – away with all that complicated system of spheres and epicycles! – and also very much because of its revolutionary daring and intellectual challenge.

Professor Galilei loved intellectual challenges, and he despised the obstacle of ossified, conservative Aristotelian thought that hindered new ideas about natural phenomena. He could not *but* be attracted to the Copernican system. On his own initiative and without discussing the matter except with close friends, he studied the revolutionary ideas. In 1597 he announced that he was a "Copernican" in one of his rare letters to his German colleague, Johann Kepler.[7]

The new ideas were not represented in his teaching, however. Certainly Galileo must have felt himself superior to the friar-mystic Bruno and the simple miller with his unfortunate penchant for peasant philosophy. But the

Inquisition was just as much a part of his daily life too, a part he could not ignore.

The Inquisition's cumbersome bureaucracy and centralised structure meant that it was not particularly effective. The Holy Office was responsible for only a tiny fraction of the executions and downright homicide inflicted on members of minority faiths all over Europe in the 16th and 17th centuries.

But the institution was a reality no less omnipresent for all that. There is no indication that Galileo rehearsed anything but pure orthodox Ptolemaic theory from his lectern in Padua. During his lectures the Earth remained absolutely fixed as the definitive centre of the universe. A professor's responsibilities included not leading his students into heresy, not crossing that invisible – and pitifully ill-defined – boundary between science and religion.

In Florence the Grand Duke Ferdinando commissioned a huge planetarium, a model of the heavenly planets and bodies. The contrivance took five years to build. It was three metres high, made of wood completely covered with gold leaf and could be turned with the aid of a handle so that the Sun and planets moved.

But the Earth stood still in the centre of the model. The planetarium was an expression of Ferdinando's eccentric love of the spectacular, but it was also a demonstration of an absolute ruler's adherence to the prevailing astronomical and theological wisdom, and thus a discreet warning to those who thought otherwise.

But the situation for those who inclined towards the theory of the Sun in the centre, was not entirely hopeless. Copernicus' book had so far not been placed on the list of forbidden works. Of even greater importance was the clear tradition that had grown up within astronomical science for strictly distinguishing between *astronomy* and *cosmology*.

True astronomy was concerned with calculating planets' orbits, the position of stars, eclipses and that kind of thing. It could have a certain practical value, especially in navigation. As far as it went, this kind of astronomy could use many different models, provided they gave sensible results. Such "mathematical models" did not necessarily aim to represent the ultimate cosmological and physical truth about what the universe looked like. This gradually also came to apply to adjusted details of the Ptolemaic system, with its epicycles and other complications (for example, that the mathematical centre of the planets' orbits was not exactly the Earth). It was taken that this was an aid to calculation and not a real description of the world.

Viewed in this light, Copernicus' system was viable as a purely intellectual and mathematical model, without the Church needing to involve itself in the matter. And a number of experiments were made in this direction, without producing noticeably better results than the old model, as Copernicus had not been very precise in his specification.

The problem was that Copernicus himself did not regard his system as a useful aid for complicated computations. He saw it as a concrete representation of cosmological reality: the Sun *stood* still, the Earth *moved* around it. Of earlier astronomers' attempts at cobbling together a tenable geocentric model, he said contemptuously:

> "They could not discover the main thing, namely the form of the heavens and the equilibrium of its parts. On the contrary, they are like one [a painter] who, from the best models selects hands, feet, head and other limbs, all of which are of the most excellent quality, but are not drawn as the picture of one single body, and therefore turn into a monster rather than a man when they are put together."[8]

Galileo also believed that the Sun actually stood still, forming a centre for the motions of the planets and the Earth. But he did not turn it into a bone of contention just then. Instead he returned to the pendulum and the falling spheres. For here, too, in miniature, there was much to learn about how the world really works.

Lecturer and Designer

The reason that Galileo quickly became a highly respected member of the academic circle in Padua was largely due to his brilliance as a lecturer, where he displayed his acute intelligence as well as his considerable linguistic skills. He had students from the Italian states and from further afield. Some came from the highest echelons of society, like the exiled Swedish Prince, who was sent there by his uncle, King Sigismund of Poland[9]. The prince even lived under Galileo's roof for a while, and Galileo gave him lessons in Italian.

The vague boundaries between subjects meant that Galileo by no means limited himself to pure mathematics. He lectured on astronomy – but did not reveal his belief in the motion of the Earth. Instead he recited the traditional, Ptolemaic counter-arguments: birds would be left behind as the Earth revolved beneath them, objects dropped from a tower would land far away from the tower's foot.

Mathematics was a "utilitarian discipline" with many applications. Galileo even lectured on military engineering, one of the subjects his teacher

Ricci at the grand ducal court had also mastered. Galileo gave two lectures on it. The first was on "the art of fortifying cities" – the second, logically enough, on how such fortified cities were to be conquered!

Providing lodging and tuition to aristocratic students supplemented his income. Galileo was perennially short of money. He was paying dowry instalments, convent fees, music lessons and the living expenses of his mother and brother. Furthermore, he was a man in his best years, and life was not all experimenting and teaching. He soon made friends, good friends, both in Padua and in Venice.

The practical Galileo, the designer and craftsman, always ready to link theoretical calculations to empirical experience, did not hide his light under a bushel. During his first years at Padua he developed a remarkable instrument for calculating and sighting, *compasso geometrico militare*. In translation this means something like "the geometric and military compass".

It was partially based on the proportional sector, an instrument used to transfer dimensions from one scale to another. Guidobaldo del Monte had constructed one. Gradually, it became reasonably common for painters to use proportional sectors, as these more easily allowed them to find the dimensional correlation between their models and what was to appear on the canvas. It must have been a really huge proportional sector of the del Monte type that once got the painter Caravaggio arrested on a street in Rome – on the spur of the moment the officer took it to be some kind of weapon!

The other prototype Galileo used was the plumb-line and square that was inserted in cannon barrels for calculating elevation, so that the projectiles would land where they were supposed to. But his fully developed instrument had a far wider application.

The geometric and military compass is a fine piece of bronze workmanship. Its two feet are about ten inches long, and one has an integral limb that can be further extended. The feet are joined by a curved cross-piece, and at the apex where the feet are hinged, a plumb-line can be attached. Feet and cross-piece are etched with lines and scales on both sides.

The instrument is *geometric*. Galileo kept to his Euclidean roots. All calculations that can be carried out with the help of the compass are approximations, based on the comparison of parts of lines and triangles. They are founded on proportionality, not on any absolute, given unit of measurement. (There were none in existence, even the commonly used *braccio* varied from town to town.) As a unit of length was needed as a basis for the quantifying of the proportions, Galileo used the more or less private measure of the *punto*, plural *punti*, approx. 0. 94 of a millimetre.

The compass was an amazingly versatile instrument. In military use it could of course serve to measure cannon elevation, but one could also estimate distance and difference in levels with it. It could be used as an astronomical quadrant for fixing the position of stars in navigation.

Its purely geometrical functions included the calculating of inscribed and proscribed circles to polygons; but one could also use it to find the radius of a circle with the same area as various rectangular polygons – estimated only, since "squaring the circle", as we know, is one of the insoluble problems of mathematics. Most interesting of all, perhaps, was that with a given polygon, for example a square, one could easily calculate the sides of a new polygon with *n* times larger area. If one selects a suitable square, this provides a direct method of finding – or at the very least of estimating – square roots (expressed as one side of the square *n*, they can be measured in *punti*). It can therefore be claimed that Galileo's compass was the first proper mechanical calculator.

Making such an instrument required great precision and took a long time. Galileo solved that problem by employing a craftsman, an instrument maker who had worked at Venice's famous shipyard, the Arsenal. The man moved in with the professor in Padua – with his entire family on a board and lodging basis. Galileo made a little money this way. The compasses sold for five scudi, which did not give much profit once the bronze had been paid for and the instrument maker had his wages. But it was complicated enough for the user to need thorough instruction. Galileo gave private tuition in its use – for a sizeable fee: twenty scudi.

A Professor's Commitments

Venice lured the young professor. The powerful old city with its black-painted gondolas punting up and down the canals, attracted him for a variety of reasons. After the discovery of the sea route to America this lagoon region at the top of the Adriatic was, certainly, in the process of becoming a backwater as a trading and maritime hub, but by comparison with Padua, Venice was a big city. And Galileo made influential friends there. Foremost among them was the wealthy aristocrat Gianfrancesco Sagredo. Sagredo had his own palace in the city's finest quarter: its slightly oriental facade reflected in the Canal Grande.

The professor from Padua was always welcome at this palace, setting out his thoughts on the physical world and its secrets. The professional and the interested amateur not only exchanged ideas, but also small gifts – Galileo might bring some truffles, and receive a present of wine from the connoisseur Sagredo.

Galileo did not go to Venice just to renew acquaintance with influential friends. He went there also to meet women, a fact that did not raise the smallest eyebrow. Even in papal Rome the most elite courtesans were invited to lively dinners with elevated prelates and foreign emissaries. One of the most eminent of these women lived in her own apartment costing 70 scudi per annum, complete with stall and standing for visitors' carriages, and she received her clients in a bed festooned with "turquoise curtains made of raw silk from Bologna"[10] and with a bedspread of the same material.

But Galileo was lucky enough to meet a young woman in Venice with whom he could form a permanent relationship. Her name was Marina Gamba and she was only just twenty when she and the professor met.

Instead of marriage, there were frequent trips to Venice. Galileo was in his mid-thirties and well established, Marina was young, poor and needed a provider – so neither she nor her family were too scrupulous about the outward form of the liaison. Marina soon became pregnant, the professor was in the process of starting a family.

Galileo brought his Marina to Padua. He did not lodge her in his house, which was already a combination of lodgings, schoolroom and compass workshop. A professor's house was a kind of extension of the university, a gathering place for serious male students, where the notion of women (not to mention the sound of children) was completely out of place.

Galileo's family life was removed to a small house just a few minutes away. There, the couple's eldest daughter, Virginia, was born on 13 August 1600, *de fornicazione*[11], as the church register blandly states, i.e. "out of wedlock". Galileo is not mentioned there, nor in the entry for the couple's second daughter, Livia Antonia, the following year. The tone is certainly a little less harsh this time: "daughter of madonna Marina Gamba and..."[12] When Marina and Galileo had their third and last child in 1606, the church registry is even more discreet: young Vincenzio is registered as "son of madonna Marina, daughter of Andrea Gamba, and an unknown father"[13].

Naturally, there was never any doubt as to Galileo's actual paternity, nor did he ever try to conceal it. The children were named after his two sisters and his father. He also cast horoscopes for them based on the time of their births – Livia would be characterised by *probitas, simplicitas, eruditio, prudentia et humanitas*. That certainly seemed pretty promising for the child: honesty, simplicity, culture, wisdom and humanity!

So why could Galileo not simply marry his children's mother? It was not impossible – his colleague Kepler, for example, had done just that. The reasons were doubtless complex, but just as surely social and financial at root. The class system dictated that Marina was hardly suited to the circles Galileo moved in, not to mention the life he aspired to: close to a princely court. Perhaps of more direct financial consideration was the fact that she did not have any dowry to speak of. The financial side of the *contract* that also was an aspect of marriage, was missing.

If Galileo officially admitted paternity of his daughters, they would be elevated to his own social class – and that, in turn, would mean he would have to provide hefty dowries when they were ready to marry, not to mention foot the actual expenses of the weddings themselves.

The professor knew a bit about the costs families entailed. He was still struggling to pay off the dowry of his elder sister. He should by now in all conscience have been getting help from his younger brother, the musician Michelangelo, but he earned so little that he had to ask Galileo for travelling money and clothes when he was offered a position by a Polish nobleman. And as if that was not bad enough, his *other* sister, Livia, was now to marry. This was to be celebrated in a style worthy of an old and distinguished, if impecunious, Tuscan family – if no one else, his mother Giulia would ensure that standards were maintained. The wedding gown alone, of black Neapolitan velvet decorated with light blue damask, cost a small fortune. And Galileo paid.

Modern Physics Is Born

It was neither as a designer of calculators nor as a Copernican astronomer that Galileo made his pioneering mark during his years at Padua. His most important work was experiments and investigations in the realms of physics. During those eighteen years he changed the foundations of traditional physics – or, as others see it, established an entirely new science. However, remarkably few people outside Padua realised this. For various reasons he did not make his results public until well into old age, and when he did finally become famous all over Europe, it was for quite different things.

Even with the leaning tower as his laboratory, Galileo had in no way solved the problem of free fall during his time at Pisa. Now he took up the challenge once again.

Galileo's writings – both public and private – are full of attacks on the Aristotelians and their unwillingness to indulge in fresh observation and reasoning. This was clearly partly an expression of his own energetic attempts to find new and more accurate means of describing physical things.

But there was also another side to this. The status of mathematicians in the academic world was low. If he could only demonstrate, with the aid of practical experiments analysed using mathematical methods, that Aristotle's interpreters were wrong, applied mathematics and experimental physics would usurp "natural philosophy's" pre-eminent place in academia, both in terms of prestige – and pay. Galileo had personal experience of how brilliant ideas did not necessarily bring with them money and recognition. He could hardly forget that, at times, his father had to let his wool merchant in-laws keep the family.

Galileo's radical renewal sprang, nevertheless, from the Aristotelian mind set, as it was taught at the Jesuits' Collegio Romano: human reason has a basic ability to recognise and understand the objects registered by the senses. The objects are real. They have properties that can be perceived, and then "further processed" according to logical rules. These logical concepts are also real (if not in exactly the same way as the physical objects).

This is the philosophical foundation of Galileo's subsequent increasing cock-suredness: there *is* a definite route to knowledge. The world exists independently of us, it is "just" a matter of understanding it correctly.

There was one fundamental problem: if we only perceive individual objects, and these are subject to all kinds of changes, how, on that basis, can we say anything definite about the characteristics common to *all* such objects – for example, falling bodies? The answer to this question is crucial to all experimentation. Early on Galileo realised that the solution was to sift out the individual and random from the particular to arrive at the general.

His experiments at Pisa had taught him that spheres of the same size but different weights fall at roughly the same speed. The difference between an iron ball and a wooden one was so small that he believed it could be explained by the buoyancy of air. But he had also realised that it was practically impossible to *measure* distances and times in free fall. The balls simply fell too fast. But it was not actually "free fall" that he was interested in, but rather what Aristotle had called "natural motion", i.e. movement that had no visible outward cause, no hand that pushed or horse that pulled.

At Padua Galileo got the epoch-making idea of using inclined planes instead. A ball on an inclined plane still moves "of its own accord", but not so quickly. Furthermore, the observer can alter the incline and see how the speed alters.

The technique of making many, comparable observations of a phenomenon to enable an underlying connection to be drawn, was not new. This had been the working method of *astronomy* since antiquity. In Galileo's time, astronomical observations were being made with greater accuracy than ever before, principally by the eccentric and despotic Danish aristocrat Tycho Brahe on the island of Hven in Øresund. The difference – which many Aristotelians would have found insurmountable – was that Brahe observed naturally occurring phenomena. Galileo wanted to arrange the "phenomena" himself, purely for the purposes of observing them.

Another important inspiration for experimentation was Galileo's experience of music. The daily routine of tuning a lute so that its sound was pure, was another sort of experimental trial and error: one had to put more or less tension on the strings, until they fell into an underlying and mathematically describable pattern.

Presumably Galileo's first inclined planes were rigged up with what looked like a tribute to his father: a copy of the finger-board of a stringed instrument, with thin, moveable bands or strings running across it. By altering the distance between these bands and listening for the click as the sphere rolled over them, it was possible for him to gain an insight into the relationship between time and the distance the ball rolled.

The first big problem he encountered, was to measure time accurately. Presumably he first tried to do this by *singing*. It was not as absurd as it may sound. A trained and skilled musician has a "metronomic" feel for the length of the sub-divided beat.

But neither the finger-board bands nor the rhythmic song were completely satisfactory. The bands disturbed the evenness of the ball's rolling movement, and singing was undeniably somewhat impracticable and imprecise. Galileo worked at getting the groove that the balls ran in as smooth and even as humanly possible. Then he also had the idea of measuring time with a sort of water clock – by simply allowing water to flow from one container, through a thin pipe and into another. If the water flow was constant, he could get a measure of how long had elapsed by weighing the water in the receptacle. The excruciating accuracy that characterised Galileo as a practical man and experimenter was visible in the way he also estimated the weight of water that remained on the walls of the container!

Galileo wanted to find out how the speed of the balls varied over distance and time. But he was operating within a Euclidean, geometrically influenced mathematical framework. In other words, he was not much interested in pure numbers. Instead, he attempted to discover the *proportions* between various stages. He was a stranger to the new algebra and he did not use decimals, only vulgar fractions. Decimals were on the way in, but it is possible Galileo did not consider the system soundly enough based in logic to be used in work that was to provide one hundred percent logically valid conclusions.

One basic difficulty in analysing the relationship between distance and time for rolling balls, he discovered, was that their speed altered the *whole time*. He had therefore surmounted the false conclusion from the Pisan period, that any falling (or rolling) body will eventually achieve a constant velocity. (In practical free fall experiments in air, increasing air resistance will eventually slow the object down so much that its speed after a certain time will become roughly constant. Otherwise it would be imprudent, for instance, to do a parachute jump.)

The very concept of "velocity" itself was not easy to grasp. Velocity equals "distance divided by time" – but what happened to the distance when he made the time interval smaller and smaller and finally asked about speed at *this* instant or at *that* point and there was no distance to divide nor time to divide it by? What did "velocity at a given point" actually *mean*?

The mathematical solution is found in the development of differential calculus, a development to which Galileo contributed, but which was outside his sphere of interests. In the absence of this aid, Galileo's concepts of velocity had been linked to completed movements over a certain distance, rather than to points. In the first instance he was content to measure how far down his inclined plane his spheres got if he increased the time they were allowed to roll. He had to keep to average measurements (distance divided by time), but he could study how much the average velocity altered over a given period. He was not, therefore, able to calculate the *continual* change in velocity, which is the real key to understanding this type of motion.

As he had hoped, his measurements revealed a rule. If the average velocity during the first unit of time was 1, it rose in the second unit to 3, in the third to 5 and so on. Using the arbitrary units "second" and "foot", the arrangement was as follows:

After 1 second 1 foot covered	average velocity first sec.	= 1 (foot/second)
After 2 seconds 3 more feet covered	average velocity second sec.	= 3 (f/s)
After 3 seconds 5 more feet covered	average velocity third sec.	= 5 (f/s)

Contented, Galileo could conclude that he had found a rule, a proportionality – if somewhat cumbersome – that concerned the increase in average speed, which was clearly proportional with the progression of odd numbers. If, on the other hand, he had *added up* the distances and looked at the total distance from the start, he would have been within a whisker of a fundamentally important, simple and general law.

But that was to come later. The most important result of the inclined plane experiments on this occasion was that velocity constantly increased as the sphere rolled. There was no "given speed" that a body would naturally reach. This hardly dampened Galileo's belief in his experimental method: he had clearly shown that here, too, Aristotle had made an elementary mistake.

A New Star in an Unchanging Sky?

In October 1604 a completely new star suddenly appeared in the constellation of the Serpent-Bearer. The star was seen across the whole of Europe, and aroused a great deal of interest. At that time the public was practically obsessed with interpreting signs in the sky, and other places. Naturally, the star was viewed as a bad omen on the whole, because people were used to war, famine and disease.

New stars appearing in the sky was not a completely unknown phenomenon. They were labelled *stella nova* ("new star") or simply *nova*. The nova of 1604 was in fact what we now call a *supernova*, a very rare stellar catastrophe which for a short period increases the light output of the exploding star a billion times or more. It was the German astronomer Johann Kepler in Prague who first noticed the phenomenon – and so the 1604 nova is known as "Kepler's nova" in consequence, and is the most recent supernova registered in the Milky Way.

Neither Kepler, Galileo nor anyone else had the slightest explanation for how the nova had come about. What they were able to do, though, was to say something about how far away it was. And this was a question of the greatest astronomical, philosophical – and therefore also religious – interest.

Kepler, mathematician to the Imperial court, wrote a book – *About the New Star* – which was largely concerned with astrological speculations. The more rationalistic Professor Galilei gave three lectures on the subject. But they both shared the same opinion about its remoteness.

The key word was *parallax*, or the angle that can be measured when one observes an object from two different points. Naturally, the greater the distance between the observations, the greater the angle. But also, the closer the object is to the observer, the greater the angle if one moves and observes it from another place. (If something is close enough, we can clearly register parallax simply by looking at it with one eye and then the other.) Or the reverse: if one looks at a star from two different places, and cannot measure *any* change in the angle of vision, it must be extremely far away, at a distance that is of quite another order of magnitude to the distance between the observation points.

Galileo did not travel about observing the nova, but both he and Kepler could easily compare data from observations all over Europe. And on one point they agreed: there was no measurable parallax. In other words, the nova was very far away – considerably further away than the moon.

This view was – to put it mildly – controversial.

The reason must once more be sought in the Aristotelian-Ptolemaic system and the theological interpretations of it. In Aristotle there is a clear distinction between what is found under the Moon (more accurately: what is found within the sphere the Moon is attached to and which revolves around the Earth), and what is further away: stars, planets and the heavenly spheres pertaining to them.

Under the Moon – in the *sublunar* zone – all was composed of the four elements: earth, air, water and fire. Here mutability and transition reigned: seasons shifted, plants grew, bloomed and withered, people were born and died, balls fell heavily to the ground if they were dropped from towers. Beyond the Moon, however, quite different natural laws applied. Everything there was made up of one single element – *ether* or *quintessence*. This had no weight (otherwise everything would have fallen down on to the immobile Earth, the centre of the universe), and the only change or movement that took place there was the "natural motion" of the spheres, in perfect circles around the Earth. By contrast, all natural movement below the Moon is straight, as a ball falls, or raindrops fall from the clouds.

It is clear that this notion had profound theological implications. Eternal perfection reigned in the heavens; earthly existence was, on the other hand, characterised by temporal frailty and change.

So by definition a "new star" could not be a star at all – as in that case a change must have occurred in the heavenly sphere where the fixed stars belonged. The nova must be some kind of natural phenomenon in the space between the Earth's surface and the Moon – in the same category as the northern lights, or the clouds for that matter.

If Galileo and Kepler were to be believed, Aristotle had made an elementary mistake on this point as well: heaven was not perfect and immutable. If the new star was not a direct argument in favour of Copernicus, it certainly put another question mark against accepted wisdom.

For Galileo, perhaps the most important result of the nova was that he had to apply himself seriously to astronomy, a corner of the "curriculum" he had not studied in depth up to that point. But he was certainly aware that the parallax question could also be turned into a serious argument *against* the Copernican theory of the Sun in the centre and the Earth in orbit, presumably the best scientific argument the Church and the defenders of tradition had.

If the Earth really orbits the Sun, said the sceptics, it must move an enormous distance in the course of the year. So, if we observe a star in the spring, and make the same observation in the autumn, the Earth will, in the meantime, have moved through space to a point diametrically opposite on its orbit, a distance many, many times greater than any we can measure on the Earth's surface. So why can we still not measure any parallax for that star? ("The greater the distance between the observations, the greater the angle.")

Copernicus had himself answered this objection. The parallax is there, but because the stars are so very far away, even in comparison to the Earth's orbit round the Sun, it is almost unmeasurably small. But this undeniably had the incontrovertible feel of the *ad hoc* argument. Anything can be proved if one can postulate data at random. (The argument was in fact *right*, but stellar parallax was first measured two centuries later, in 1838.)

Drawing Close to a Court

No jubilees were celebrated in Florence, but even there the year 1600 was an eventful one. Grand Duke Ferdinando had plenty of excuse to create the kind of lavish entertainments that he loved. The greatest of these occurred on one of the most glorious occasions in his family's history: Ferdinando's niece, Maria de' Medici, was marrying the French King Henri IV of Navarre. True, the ceremony in Florence's Santa Maria del Fiore Cathedral was conducted with his proxy, but this in no way dampened the festivities. There were horse races, jousts, processions and fireworks – and great musical performances. Galileo's close friend, the painter Cigoli, had links to the inner clique *La Camerata* as a lutenist, and there is strong evidence to suggest that he also designed what we would nowadays call the scenery for *Eurydice*. It was the world's first operatic performance and it was produced at the court that year.

Cigoli had other important commissions in his home city, both as an architect and a painter, where he was a key figure in the break with Bronzino's cold "Mannerist" style. Cigoli's was the ultimate court style, the baroque, which grew up within music and pictorial art during the years around 1600.

From his professorial chair in Padua, Galileo kept a close eye on events in Florence, and not just on the arts. Cigoli and others kept him informed on matters large and small. The court clearly needed expertise of many types, and time would show that certain "performances" were so grand that they required people with a knowledge of practical engineering and physics.

But it is likely that, through his various channels, he followed the fortunes of Grand Duke Ferdinando's eldest son, Prince Cosimo, of whom he harboured great hopes. In 1601 Galileo received a letter from a friend and colleague in Pisa, Professor Mercuriale, who was also physician to the Medici family. As a friendly hint to a talented son of the city living in exile, Mercuriale mentioned the Prince's future to Galileo: the boy was to succeed to the grand ducal seat one day – and in the meantime, might he not be wanting a good mathematics teacher?

Galileo's social position prevented him from applying to the court directly with any such enquiry. He had to let intermediaries look into the matter, and not until 1605 did he feel secure enough to approach the then 15-year old Prince directly in a particularly obsequious letter:

> "I have, until now, made certain to send the necessary marks of my esteem through my most trusted friends and benefactors, because I did not deem it seemly – in leaving the obscurity of night – to show myself directly before you and to look into eyes that have the clearest light from the rising sun in them, without first having prepared and fortified myself with that light's reflection."[14]

The letter brought results. The summer of 1605 was spent by Galileo as private tutor to Cosimo in the villa in Pratolino outside Florence, where once the Grand Duke Francesco had retired to nefarious studies with his Bianca.

Grand ducal family life was idyllic now, compared to conditions under Francesco. Christina of Lorraine was a pious and deeply religious woman who eventually had nine children with her Grand Duke, eight of whom survived: four daughters and four sons. The latter Galileo would later exploit in a particularly propitious and elegant manner.

Galileo did well with young Cosimo, who was far from untalented at mathematics. After his job was done, Galileo managed to use this first direct contact with the Medicis to obtain a much-needed wage increase at Padua. (It was true that his salary had been raised from 180 to 320 scudi in 1597.) Through his emissary in Venice, Grand Duke Ferdinando hinted that his eminent countryman, Professor Galilei, might possibly be slightly underpaid in his post. The Venetian Senate was normally sceptical about attempts to influence them by foreign potentates, but clearly this one was not taken amiss as Galileo's salary was increased to 520 scudi.

After this successful summer, Galileo planned the next push towards the princely court in Florence. This was in connection with his invention, the geometric and military compass. Such an instrument – naturally in a specially constructed version made out of precious metals – ought to prove a timely gift to an up-and-coming potential military commander like young Cosimo, especially now that the boy had learnt enough maths to use a few of its functions. And Galileo could kill two birds with one well-aimed stone: he would print a small book that gave an introduction to the use of the compass, and thereby relieve himself of the private tuition, while still being able to earn money through the sale of the compass and the book. The book could be fittingly dedicated to Prince Cosimo, with assurances of his humble esteem:

> "If, mighty Prince, I were to attempt to record on these pages all the praise accruing to Your Highness' own merits and those of your incomparable family, I would be forced into so voluminous an account that this preface would far exceed the length of the remainder of the text."

It is against this background we must view Galileo's excessive fury towards Baldassare Capra, the author of a pirate edition (in Latin) of *The Workings of the Geometric and Military Compass*. Galileo's biographers have often been somewhat alarmed at the temperament Galileo displayed in this inconsequential matter. Galileo was unusually proud and had an excitable disposition to match. And certainly, such a pirate edition might have financial consequences. He took the matter to law and the court found fully in his favour: Capra's book was impounded. But Galileo was not satisfied. He had a pamphlet printed in which his adversary was told in no uncertain terms that he was "a malicious enemy of mine and of all mankind" and his writings are called "the poison from this evil lizard"[15], to mention but two of the epithets that bombarded the unfortunate mathematician and petty swindler.

As far as Galileo was concerned it was not simply his abstract "honour" that was at stake, although that was important enough. Capra had sullied his gift to the future Grand Duke, meddled with the well considered strategy that would carry him towards the Florentine court.

The Grand Duchess herself made sure that Galileo was invited when young Cosimo was married in 1608. The marriage was yet another dynastic triumph for the Medici family. The bride, the Austrian Archduchess Maria Maddalena, was the sister of Ferdinand of Habsburg, later to become Emperor Ferdinand II. The celebrations for this wedding exceeded anything that even Florence had become accustomed to. The river Arno was turned into a "stage", with tribunes along its banks. On it was performed a piece about Jason and the hunt for the golden fleece, complete with giant dolphins, menacing lobsters and a fire-spewing Hydra.

It was to be Grand Duke Ferdinando's final glittering show. In January 1609 Galileo got a letter in Padua from the Grand Duchess Christina requesting him to cast a horoscope for Ferdinando, as he had become seriously ill. Obediently, Galileo gazed into the stars but without his usual perspicacity for, despite predicting many years' of happy life for the great man, Ferdinando died a mere three weeks later.

With Ferdinando's passing, the last Medici of any consequence had gone. The 19-year-old Grand Duke Cosimo II had certainly inherited his father's feeling for magnificent processions, but – despite Galileo's words and all the praise lavished on his merits – very little of Ferdinando's calculating intelligence and political sagacity.

But for Galileo the succession represented a wonderful chance. Ferdinando's accession to the title had opened the door to the academic world back in 1587. Now the Grand Duke's death just happened to provide a golden opportunity to get out of that world and into another, even more promising one.

The Balls Fall into Place

After his teaching, family life and practical tasks, it was his experiments that absorbed Galileo most. And because he could get no further with his inclined plane, he began once more to look at the motion of the pendulum, a subject that had interested him since his student days at Pisa.

He knew that the time a pendulum took in its oscillations was constant and not related to the distance it travelled – provided its movements were small. But the rate at which it swung had a clear correlation to the *length* of the pendulum, although no one knew what that correlation was.

It was relatively easy to measure the time of constant pendulum swings with some accuracy, since he could time *many* swings and divide by their number. Galileo realised that the movement of a pendulum is also a kind of fall – a "natural motion" not dictated by any outside force. (From a modern perspective this is not true because the pendulum is affected by the force of gravity. But Galileo knew nothing of this.)

And so he began to time the oscillations of pendulums of different lengths. When his work room got too small, he went up to the top floors of the university and hung the pendulum out of the window – the longest one he tested, was well over nine metres long. Time was still measured by weighing the amount of water that had run into a container.

After a while he found a relationship between oscillation time and length. With an appropriate choice of time unit (he called it the *tempo*, literally "time"), he could draw a simple proportion on the geometrical model in which the oscillation time was the *mean proportion* between 2 and the pendulum length. Written in modern form with time = T and pendulum length = l it is:

$$\frac{2}{T} = \frac{T}{l}$$

As the interested reader can easily work out, this is the same as saying that the pendulum's oscillation time is proportional to the square root of the pendulum's length.

Thereafter he returned to his falling bodies to investigate what happened to them when they covered the same distance as the pendulum's length. Now he knew what he was looking for. To his great joy he discovered that it seemed that *falling time*, too, was proportional to the square root of the length of the fall. (The absolute figures were, of course, different to the pendulum's oscillation time.)

So here was the key to the precise description of "natural motion". All that remained, was to take out his inclined planes again and repeat the experiments where it was easier to vary and measure. It worked here, too – the time was proportional to the square root of the distance. When Galileo looked back at his old proportions for average velocity over various times, he could see that this law had been staring him in the face in his figures: if he measured the combined *length* a ball had rolled, the distance from its starting point was proportional to the square of the *time* it had taken!

It was that simple – and that difficult.

So much of the old mind set had to be shut out, so much precision and orderliness had to be invested in the experiments, so much reflection had to be expended over the bare figures to find out the relationship that linked them. Galileo's finding became known as the "law of falling bodies" and forms the basis of all modern teaching on motion, the branch of physics known as *kinematics*. In fact, this precise mathematical description of an idealised physical motion – movement by constant acceleration – is really the very foundation of modern physics.

Galileo formulated his law as a proportion, on the geometrical pattern, and not in the modern form we know from school textbooks:

$$s = \frac{1}{2} a t^2$$

where s is distance, a acceleration and t time.

It is highly doubtful if an arithmetical operation such as "squaring time" would have had any meaning for Galileo. Nor does it for us – at least we never speak of "square seconds" or "square years". Galileo brought the natural sciences to a decisive watershed as regards people's understanding of the world about them: scientific description of natural forces becomes clothed in mathematical language and moves decisively away from *common sense* and the everyday observations any one of us can make. In short, the world is far stranger than it appears at first glance.

This realisation was a fundamental prerequisite for the breakthrough of Copernican ideas: that one could *calculate* that the Earth was revolving at speed round the Sun, even though it certainly did not seem to be. Galileo fired the starting gun in this development with his statement:

> "Philosophy is written in this grand book – I mean the universe – which stands continuously open to our gaze, but it cannot be understood unless one first learns to comprehend the language and intercept the characters in which it is written. It is written in the language of mathematics, and its characters are triangles, circles and other geometrical figures (...)"[16]

The Roman Style

One of Galileo's closest friends in Venice was the Servite monk Paolo Sarpi. Brother Sarpi was an extremely scholarly man who had occupied an elevated position within his order at its headquarters in Rome, where he had been on good terms with Pope Sixtus V and above all with the powerful Jesuit, Robert Bellarmine. When the heads of the order wanted to reform their cloisters, Sarpi was chosen for the job. He was sent northwards from Rome as a highly trusted and respected cleric.

Sarpi was, in reality, a man full of doubts, influenced by the Reformation and the new ideas of his age. He wriggled out of the assignment, settled in Venice and offered his services to the republic, but remained within the Servite Order. In this way he led a kind of double existence. He wore a mask that protected him from the wrath of the Inquisition. Behind it, he lived a reserved and cautious life with the equivocation and sceptical ambivalence he felt towards all accepted truths, whether in religion, politics or science. Amongst the things he doubted was the Holy Trinity, and he believed Jesus was a prophet, not the son of God. He discussed such views with rabbis in Venice, an intellectual exercise that was far from safe.

Sarpi and Galileo were naturally drawn to each other. Theological questions certainly did not interest Galileo, but Sarpi's sceptical inquisitiveness was all-encompassing. The two discussed cosmology, mechanics, kinematics and the theory of heat – the latter resulting in a rather imprecise instrument for measuring temperature, the "thermoscope".

Relations between Venice and Rome were strained. The Republic defended its independence, while the papacy was worried that heretical ideas from the north might creep in through Venice, and tried to assert its supremacy in all matters connected with religious life.

The crisis came when two priests were arrested in Venice in 1605 and accused of murder. The church demanded that, following the normal rules, they be handed over to their clerical superiors, who would then look into the matter. The Venetian Senate refused to hand them over and instead arraigned the priests before a secular court. The church regarded this as a serious attack on its privileges.

In Rome, the jurist and theologian Camillo Borghese had just been elected pope under the name Paul V. His personal lifestyle was simple and modest – but he had very definite ideas about the absolute authority of the papacy. The very day after his election he ordered the immediate decapitation of a writer from Cremona, whose offence had been to compare a former pope with the Roman Emperor Tiberius.

Pope Paul was furious at Venice's treatment of the two priests, and mobilised his best ecclesiastical lawyers to draft a recommendation, a legal basis for action against the Republic. This work was handed over to a group of Jesuit experts, led by Cardinal Robert Bellarmine, Sarpi's old friend, and the man who had played a decisive role in the case against Giordano Bruno.

Based on this recommendation, the Pope brought out his greatest weapon, apart from open warfare: in April 1606 he placed an interdict on Venice. He forbade all ecclesiastical services to the Venetians, like mass, the giving of communion and other sacraments, including Christian burial. This meant that anyone who died in Venice, had to spend eternity in a Dantesque hell.

During the Bruno affair the Venetian authorities had not bent over backwards to try to save the sceptical friar from extradition. But now it was no longer a question of a friar's fate, or that of a couple of priests – it was a question of power, of just how much legal sovereignty an independent state really had in relation to the Church.

So the Senate struck back hard and fast. It ordered all priests in the Venetian region to regard Rome's interdict as invalid, it expelled all Jesuits from Venetian soil and drafted its own legal counter-recommendation which concluded that the Pope's bull of excommunication was "not in keeping with any natural reason whatever, and in contravention of the teachings of Holy Writ, of the Church Fathers' doctrines and of the holy canonical writings". The interdict was therefore "not only illegal and unwarranted, but also void and without force of any kind"[17].

This recommendation – and here one can really talk about "spitting in Rome's eye" – was formulated by Paolo Sarpi. The friar represented a new era in the political field. He was defending the modern, secular state which can arrive at independent decisions on an impartial legal, rather than a theological, basis.

In so doing Brother Paolo Sarpi had stepped forward and shown that, behind the mask, he was a very courageous man. But when the crisis developed to the point of war, he was ordered to present himself at Rome. He very wisely refrained from making the journey.

The interdict was rescinded after a year, after a compromise that satisfied neither of the parties. But there were some who remembered Sarpi's involvement.

Venice's labyrinthine network of narrow, dark alleys should have been an ideal place for assassination. But the two armed men who attacked Sarpi one autumn evening in 1607, did not manage to complete the job. They left the friar bleeding profusely, with a stab wound through his cheek, but not dead. Whether it was the Pope himself or the even more hurt and enraged Bellarmine who was behind the murder attempt, has never been firmly established.

Sarpi had his own suspicions. "I recognise the Roman curia's *stilus*," he said. *Stilus* can mean both "style" and "sharp instrument". The attackers got away.

The Senate realised that Sarpi needed better protection. A papal spy who was caught in 1610 while trying to ingratiate himself with Sarpi's secretary, was immediately sentenced to drowning in the lagoon; a punishment he avoided by admitting and documenting that he had been sent by the Pope. Gradually Sarpi found peace enough to begin his magnum opus: the history of the Council of Trent, a critical historical review of the very basis of the Counter-reformation.

Galileo followed his friend's deeds at a certain respectful distance. Political and legal matters were not within the ambit of his interests, nor in fact was Venice's independence, the more so since he had plans to leave the Republic as soon as an opportunity presented itself. He kept in contact with Sarpi, but certainly had no interest in clashing with papal authority. He might have good use for his contacts in Rome. Above all he did not share the Venetian repugnance of Jesuits, who after all was said and done included several of Italy's finest mathematicians in their ranks.

He was even less willing to fall out with the deeply religious Grand Duchess Christina, who still wielded a lot of influence at the court in the Palazzo Pitti. Because now, the chance he had been waiting for had arrived.

The Tube with the Long Perspective

Paolo Sarpi had contacts all over Europe. At some point over Christmas 1608 he heard rumours that a spectacle-maker in the Netherlands had constructed a long tube. If one looked through it everything seemed to get closer and appear larger.

The Servite monk was not especially interested in it himself, but when the rumours got louder, and one considered the military implications such an instrument might have, he mentioned the phenomenon to his friend Galileo, one evening in June 1609 when the professor was on a visit to Venice.

The inventor of the geometric and military compass realised immediately that here was a completely new opportunity. Whereas the compass could merely calculate the distance to the enemy's emplacements, this new instrument could *display* them! The "spyglass" was, at the same time, a practical and intellectual challenge and possibly a financial godsend.

Galileo grasped that the secret was connected with lenses, of the sort spectacle-makers used. As Venice was a glass manufacturing centre, it was easy to get hold of a suitable selection of different lenses. He then travelled straight home to Padua and began work.

He was not acquainted with any practical theory of refraction, so he felt his way forward by trial and error. The instrument was shaped like a tube, so it was reasonable to assume there was a lens at each end, with a certain distance in between. But what type of lenses? There were curved lenses of two types, convex and concave, and one that was flat on one side and curved on the other.

After a day's experimenting Galileo had a primitive telescope. He placed a plano-convex and a plano-concave lens in a tube and got an out-of-focus image of distant objects, magnified some three or four times. Proudly he returned to Venice by water and showed the result to Sarpi and other friends: whatever foreign inventors could do, the Republic's own professionals could easily copy and surpass it!

His friends kept the telescope. Galileo returned to Padua to carry on the work.

He was fully aware that the instrument was not good enough, and that improved versions would certainly be made by others. So he attempted to understand the theory behind its operation, and at the same time he learnt to polish glass.

Meanwhile Sarpi and the others exerted themselves in Venice. They contacted the Senate, demonstrated and boasted about the new invention. Galileo was praised and promised better pay, and he worked on conscientiously with his lenses. By August everything was ready for the big demonstration.

Once again Galileo was clambering to the top of a belfry, this time the slender and lovely detached *campanile* in Venice's famous Piazza San Marco.

With him, up the winding, stairless climb went the leading men of the Republic, senators and others. He carried the instrument himself; he called it a *cannocchiale*. The tube felt heavy in his hands – it was made of lead and covered with a crimson cotton fabric. It was about sixty centimetres long and fairly narrow.

The day was crystal clear. From the top of the tower one hundred metres above San Marco, the view stretched in all directions.

Each of those invited tried it in turn. They placed the tube to one eye, closed the other, and pointed it over the lagoon.

One of the men pointed the tube to the north, towards the glass-manufacturing island of Murano, about a mile off. He had a little trouble locating the church of San Giacomo within the small field of vision but, once found, he could clearly make out the people who were going in and out of the church door. A little way off a gondola was tied up on the Glass-makers' Canal, and people were disembarking.

Another turned it to the south-west and followed the coastline with the glass. He saw something that had to be Fusina, the place where the canal from Padua disgorged. And so they went on, round the horizon – until one senator caught sight of something that surely was a distant cupola or campanile far inland. He pointed, they discussed the direction, several others took a look – and then they agreed: it must be Santa Guistinia at Padua!

Galileo had enabled them to look out across the plain from Venice's highest tower and see all the way to the city where he lived.

Venice looked outwards, towards the Adriatic. The sea brought riches to the Republic – but also threats. None of the senators was in any doubt about the implications of Galileo's *cannocchiale*: an enemy ship could be observed several hours sooner, and the defenders could get an idea of its size and armament.

After the convention of the times, Galileo had *presented* his invention to Venice and its Senate. He was an academic, not an artisan who sold his services. It was up to the administrators to show how much they appreciated his gift. He could hardly complain about their response: the professorship at Padua was his for life, and his salary was almost doubled to the handsome, round sum of 1,000 scudi per annum. It was taken as read that he would not divulge the workings of the magic tube to anyone else.

But here the Senate made a miscalculation. A sceptical Tuscan agent in Venice reported to the Grand Duke's court: "It is said that in France and in other places the secret is already well known, and can be bought for a small sum." And then, just a couple of weeks later: "Signor Galileo's secret, or 'the tube with the long perspective' is now being sold here publicly by a certain Frenchman..." The agent had, however, to admit that Galileo's telescopes were far superior, a supremacy which – thanks to the professor's practical ability and technical insight – was to continue for several years.

A New World

The telescope with the crimson covering magnified nine times. Galileo made many of them during the course of the autumn. He did not feel constrained by his promise to the Senate to keep the invention secret – the more so since basically it was not his own invention – but he had to admit that the products were of variable quality depending on how well he succeeded with the lenses.

The objective lenses in the first telescopes were small. Galileo discovered that he could cut them more accurately if he made them larger, but this affected the sharpness of the image. Then he began to stop down the objective, to cover most of it so that only a small opening was left. This was an improvement, and so were his experiments at fashioning the tube so that it could be pulled out or pushed in.

Although Galileo also tried to describe the theory of "perspective" and "refraction", it was his practical experiments that gradually improved the telescope. By early autumn he had an example that would magnify twenty times.

As soon as the nights at Padua grew long and dark, he took the next step. He raised his *cannocchiale* and pointed it to the sky.

Galileo still had not gone public with his views on cosmology. As far as the Copernican system was concerned, he still lived and taught according to a Sarpi maxim:

> "Your innermost thoughts should be guided by reason, but you should act and speak only as others do"[18].

But he had a definite feeling that the general consensus about astronomical truths was basically just as flawed as he had shown the "truth" about bodies in free fall to be. Indeed, he suspected, though he could not prove it, that the spherically shaped bodies in the sky in some way adhered to the same simple laws as the earthly balls he had carefully timed on hundreds of occasions rolling down his inclined planes.

The telescope now afforded him a golden opportunity to observe heavenly phenomena that no one else had yet seen. But competitors would soon have similar instruments in their hands, so he had to make the best use of his head start.

The Moon was the obvious and easiest object to observe. From the end of November he took regular observations which provided a totally new and startling picture. Instead of the smooth spherical surface that the textbooks described, the Moon's surface was clearly disturbed and rough, with valleys, mountains and craters. No circular, Aristotelian perfection there.

A new world opened up as he lifted the telescope yet higher: the broad band of the Milky Way dissolved into stars, myriads of unknown stars! They had never been seen by Ptolemy or anyone else, were not marked on any celestial chart, had never been accorded any astrological importance. But they existed.

All this was important enough. But his great, ground-breaking discovery began on the evening of 7 January 1610.

Galileo, a professor of mathematics, was no practised astronomer of the Tycho Brahe school accustomed to making painstakingly accurate reckonings. But that evening he tried to locate Jupiter whose aspect was then favourable. As the objective lens had had to be stopped down using a piece of cardboard with a hole in it, the angle of vision was minimal. When the professor finally found the planet, he noticed that it appeared as a tiny flat disc, and not just one point of light. In addition he saw three small, unknown stars which were directly in line with Jupiter and lying close to the planet. Two of them were due east of Jupiter, the third further west.

The sighting was odd, but Galileo had already found that his telescope picked out many new and unknown stars. And so he calmly noted down the new find, pleased that he had made yet another discovery that would astound the astronomical establishment and bring him yet more kudos – and, perhaps, in the long run, an even greater income than the Venetian Senate's thousand scudi.

The very next evening, 8 January, found him ready to make more observations of the moving planet Jupiter, which according to the calculations and tables should now have shifted its position in the night sky a little in relation to the fixed stars.

As Jupiter, seen from the Earth, should have been moving *westwards* just then, Galileo expected the planet to have passed the third of the newly discovered stars such that all three would now be *east* of the planet. But when at last he managed to find Jupiter in his small field of vision, there was nothing to be seen on the eastern side. The three small stars all sat neatly on its *western* side.

Galileo sketched the constellation on a piece of paper and compared it with his notes from the evening before. There was no room for doubt: Jupiter was not behaving the way it was supposed to. He had no idea why. But as most apparent mysteries are eventually found to have simple explanations, he imagined it might have to do with some error in the tables. For most of the year, Jupiter's observed orbit moved *eastwards* in relation to the fixed stars. If this held good for these January days as well there would be nothing to explain at all.

Even so, he was in some anticipation about what he would see on the next evening. But 9 January was a mild and overcast day in Padua. The sky was a uniform grey, not the slightest glimpse of any star penetrated the cloud cover.

10 January 1610 was, by contrast, a cool clear day and the good weather lasted until nightfall. As soon as it got dark, Galileo set up his telescope. He fixed it to a tripod – without support it was difficult to hold the instrument still enough when one was looking at distant objects. Then he cleaned the lenses and turned the telescope to the point in the sky where Jupiter should now be.

What he saw was enough to convince him that he was on the track of something incomprehensible, something entirely new and unknown.

Only two of the stars were now visible, but this evening both were back where he had seen them three days earlier, in a line *east* of Jupiter.

Perhaps Galileo Galilei's principle characteristic as a scientist was his uncanny ability to draw fast, almost intuitive conclusions from a limited number of observations. He once wrote that this was the way God himself reasoned: "...immediate conclusions, without transitions, is what characterises God's mind." (But when it was Galileo and not God who was making the mental leaps, the conclusions were not *always* correct.)

He sketched the three heavenly bodies once again and pondered the phenomenon. The tables of Jupiter's movements had no bearing on the matter. Regardless of any failings there, planets always moved more or less evenly eastwards or westwards, they did not jump backwards and forwards from one evening to the next.

There was obviously a possibility that, from one observation to the next, he had mixed up the small new stars with other stars near by. But no matter how carefully he searched the sky around Jupiter, he found no other stars. And the only reasonable explanation for the apparent absence of one of them, was that for the time being, it was hidden behind Jupiter.

Only one explanation was left: the strange phenomenon had nothing to do with Jupiter's orbit. *It must be connected with a motion in the small stars themselves!*

The problem was that according to all astronomical wisdom from the Greeks to Tycho Brahe, the fixed stars stood still – that was why they were called fixed stars. Only planets moved, and then only in their fixed orbits and not in this confusing way.

Even that night Galileo had suspicions about what he was looking at. But on the next evening, 11 January, when the small stars were displaying yet another configuration, he was certain. The matter was, he wrote, "as clear as day": the small points of light were neither stars nor planets. They were – no matter how ridiculous it sounded – *moons.*

In contrast to the vast majority of his contemporaries, Professor Galilei certainly despised astrology. But he could read a portent in the sky when he saw one. He knew very well what these satellites – two evenings later he discovered there were actually four of them – heralded. They presaged not just a revolution in astronomy, but if he acted swiftly and surely, they might also bring about a dramatic change in his own fortunes.

Jupiter's Sons

In Florence's Palazzo Pitti, the twenty-year-old Grand Duke Cosimo II pursued a life of courtliness and fashion, while his mother and wife dedicated most of their time to religion. The ruling family distanced itself more and more from the realities of political everyday life. In consequence the *symbols* of power and their foundation assumed a greater importance.

Galileo, with his good contacts at the court, was well aware of this. As early as 1608 he had attempted to persuade the Grand Duchess Christina that his scientific insight could be turned to some profit in an advanced use of symbolism. He had written to her and suggested that a medal be struck to commemorate her son's marriage to Maria Maddalena. The central emblem of the medal was to be a globe-shaped magnetic ironstone that was attracting small pieces of iron from every side. The magnet's attractive power – under the motto *Vim Facit Amor*, "Love creates strength" – was to symbolise his subjects' unconquerable yearning for their ruler, but also the irresistible power that he radiated.

No medal materialised, presumably the symbolism was a bit too high-flown both for Cosimo and for the average courtier. But there was nothing wrong with the idea. And it could even be applied to Jupiter himself.

Galileo's feverish scientific and career-orientated efforts during the spring of 1610 must be viewed as two sides of the same coin. If he was to create room for a modern, non-Aristotelian natural science, based on experiment, observation and mathematical analysis, that science had to have status. And status was linked to the exponent's position within society – not to collegial approbation from within the ivory towers of academia.

The name Cosimo was often linked to the *cosmos*. Cosimo I had liked to represent himself as the fulfiller of Florence's predestined fate, guaranteed by the cosmos through the suitable horoscopes he ensured were cast. In this Medici mythology, Jupiter – both the god and the planet – played a leading role. Just as Jupiter was the chief god, the father of a divine dynasty, so Cosimo I was the founder of a line of grand dukes, a family of absolutist rulers raised high above everyday life.

Now Cosimo II had the title. One could only approach rulers by making unconditional gifts. Galileo had presented his telescope to his employers in Venice, and been rewarded. But now he had an even more spectacular present – if not up his sleeve, at least within his gift.

Only the previous autumn Galileo had shown the telescope to Cosimo II during a visit to his home city. The young ruler had been interested, and so now Galileo settled down to write a brief report to the court in Florence about his new discoveries. He soon learnt through his contacts that the nobleman was amazed and impressed at his old tutor's achievements. So were his three younger brothers.

Galileo had already begun to write a short book about his telescopic discoveries, while at the same time continuing his observations. Speed was now of the essence. On 13 February he wrote to Cosimo's "prime minister", his Secretary of State, Belisario Vinta, asking what he should call these four heavenly bodies: the *cosmic* or the *Medicean* stars?

The four satellites of Jupiter were the greatest new astronomical discovery yet made. Galileo literally wanted to dedicate them to the Medicis by giving them names and incorporating them into the family's symbolic universe.

But haste was needed. Personally, he liked the association "Cosimo-cosmos" best, and described his find using that terminology. This meant that when the book was almost ready for the printer, and the reply came from Florence, he had to do a small cutting and pasting job. Although the young Grand Duke was graciously pleased to accept that Jupiter's four satellites travelled through space in homage to him and his three brothers, his wish was that they should be named after the family.

So it was to be the Medicean stars.

Galileo's book about the discovery came out on 12 March. It was aimed at a learned European public and was written in Latin, not Tuscan. The 500 copies were sold out within the week. The title was *Sidereus Nuncius*, which could mean "Starry Messenger" as well as "Starry Message". It was the latter meaning Galileo had intended. Amongst several pages of dedication he writes to Cosimo II:

> "Scarcely have the immortal graces of your soul begun to shine forth on earth than bright stars offer themselves in the heavens which, like tongues, will speak of and celebrate your most excellent virtues for all time."[19]

The message the stars had was a tribute to the Medicis. It was impossible to say more plainly that Galileo had found scientific evidence for the family's dynastic horoscope.

But interwoven with this message there was another: Professor Galilei, Florentine, possessed all the necessary qualifications, both as regards his virtuous scientific endeavours and his mastery of the courtier's well-turned exaggerations, for the post of mathematician to the court of Florence.

Such a position could not be applied for, it was a mark of grace and favour.

As well as his copy of *The Starry Message* Cosimo II also was presented with the telescope that Galileo had made the discovery with. Now the Grand Duke could see for himself, if he was in doubt. Galileo travelled to Florence and Pisa in the Easter holidays to point out the satellites personally to the Grand Duke and his court.

The Grand Duke and the court *did* have doubts. So far, it was Galileo alone who had identified the Medicean stars. The telescope was so primitive that it required an extremely practised and skilful operator to locate the objects, four tiny pricks of light at the end of a telescope tube. Should the discovery be disproved by others, there would be little honour for Cosimo, but rather international ridicule for a conceited nobleman who accepted all acclaim indiscriminately.

The first resistance did indeed show itself fairly quickly. Not entirely unexpectedly it emanated from the University of Bologna. It was the seat of the mathematician Giovanni Magini, Galileo's old rival.

Galileo elected to visit Bologna on his way back from Florence, so that he could personally demonstrate the four new celestial bodies. It was a studiously polite meeting of fellow scholars. Magini organised a professional get together in the evening, including colleagues and students. Everyone got the chance to turn the telescope skywards.

But Magini did not, or would not, see the satellites. In addition he was far from persuaded that discoveries made with the aid of a telescope were scientifically valid. Who could say but that the new phenomena were not deceptions or chimera springing from the very construction of the apparatus?

Magini had a young Bohemian student named Martin Horky living in his house. The professor allowed his student to spearhead the attack on Galileo, a not unknown tactic, and one which Galileo himself would employ. The good Horky took up the cudgels with zeal. In a letter to Johann Kepler no less, he wrote:

> "But all acknowledged that the instrument deceived. And Galileo became silent, and on the twenty-sixth, a Monday, dejected, he took his leave from Mr. Magini very early in the morning. And he gave no thanks for the favours and the many thoughts, because, full of himself, he hawked a fable. [...] Thus the wretched Galileo left Bologna with his spyglass."[20]

Without losing any time, Horky published a small book which he called *Contra Sidereum Nuncium* – "Against the Starry Message". Admittedly he had to concede that the telescope performed wonders *in inferioribus* – "in the lower regions". But it was unusable *in superioribus* – for making observations of astronomical phenomena. In this way Horky could claim an Aristotelian basis for his criticism. But his attack was far too virulent, he accused Galileo of being an academic charlatan and compared Jupiter's satellites with attempts to find the square of the circle. This was too much for Magini who kicked the Bohemian out of his professorial house. But the rumours of Galileo's reported fiasco spread quickly. Horky sent the book to anyone who might have influence, even to Paolo Sarpi – and not least to Florence.

There it was read with the greatest interest. Patriotism was not the only trait that flourished in Florence. It was mixed with an equally intense feeling of envy and scepticism towards children of the city who stood out, something Dante could certainly have vouched for. Amongst those who fell on Horky's pamphlet were two local philosophers called Sizzi and delle Colombe. The latter name means "of the doves". The man was obviously far from pleased that the eagle Galileo was coming back to his native city.

For Cosimo II had overcome his doubts. On 10 July he appointed Galileo grand ducal mathematician and philosopher. Formally he was to become an extraordinary professor at the University of Pisa, but without any residential or teaching obligations.

The reason that Cosimo let himself be persuaded so quickly, despite the resistance, was principally due to the full, enthusiastic and unconditional support Galileo received from the most highly regarded astronomer in Europe. It was welcome – and more than a little strange considering that this astronomer, too, had not managed to make out the satellites of Jupiter through the telescopes he had at his disposal!

Johann Kepler, Imperial Mathematician

One of the most extraordinary things about Galileo's life and work is his relationship with his greatest colleague, a man seven years his junior, whom he never met and only very rarely corresponded with. The German Protestant Kepler had the keys that Galileo needed. They were moreover simply and strikingly formulated "in the language of mathematics".

Johann Kepler had been born in the small town of Weil on the Rhine, where the river forms the border between modern Germany and Switzerland. He became a teacher at a Protestant grammar school at Graz in Steiermark, in the south of what is now Austria, and where in 1596, at the age of twenty-five, he published a great work *The Cosmographical Mystery*, a sober description of the Copernican system, mixed with large doses of religiously influenced numerology.

Kepler assumed that the planets – including the Earth – revolved around the Sun. At that time there were six of them (the discovery of Uranus, Neptune and Pluto was awaiting the telescope). In the five spaces between the planets' orbits Kepler believed he could prove – with the help of a good deal of guesswork – that the five so-called *perfect bodies*, consisting of regular polyhedrons, were found.

Using this astute construction, he turned the religious argument against Copernicus on its head. He declared simply that God's ingenious plan for creation was not *refuted* by the Copernican system, but on the contrary, it was *demonstrated* in its full perfection: "Now you see how, through my endeavours, God also allows himself to be acclaimed in astronomy," he wrote to his teacher Mästlin. More sceptical colleagues were impressed, but not convinced.

Galileo also got a copy of *The Cosmographical Mystery*, and Kepler asked for his comments on it, but they never materialised, only a non-committal letter of thanks in which Galileo acknowledges that he shares the conviction that Copernicus was correct – but does not want to state the fact publicly.

Prague was the seat of the ruling Holy Roman Emperor, Rudolf II. As the power struggles raged around him, the Emperor retreated into studies of art and science. Rudolf saw to it that Tycho Brahe, who had virtually fled Denmark after complaints that he mistreated his tenants, was summoned to Prague as Imperial Mathematician.

Tycho Brahe recognised that Kepler was a genius and asked him to come to Prague, to cease his speculations and concentrate on empirical observations, the area in which Brahe himself was a passed master. Kepler did go – but not before the fateful year of 1600, and then only because he, as a Protestant, had literally been driven out of the predominantly Catholic Steiermark.

Kepler was a mystical and speculative theoretician. Brahe's strength lay in minute observation in which every assumption was checked. In his heyday in Denmark he had dispatched an expedition to Frombork merely to check that Copernicus had got precisely the correct latitude for *his* observations. Kepler and Brahe did not get on well personally, but their work changed astronomy for ever. Their first meeting took place on 4 February 1600, a date which may well be called the dawn of a new age.

Kepler, shy and sensitive, soon discovered that the rather blustering Danish aristocrat was not easy to work with. In addition, he did not much like the work he had been given – writing a pamphlet attacking one of Brahe's opponents! But their direct collaboration did not last long, because in October 1601 Brahe died suddenly and unexpectedly. The story has always been that it was court etiquette that killed him – his bladder is said to have burst because he could not rise from the Emperor's table before the Emperor himself. But in all probability he died of lead or mercury poisoning, perhaps as a result of many years' experimenting with chemicals.

Emperor Rudolf II of the Holy Roman Empire had no scruples about appointing yet another Protestant – Kepler – to the post of Imperial Mathematician after Brahe. In this role Kepler was given the task of putting Brahe's posthumous effects in order. And this would have given him access to something Brahe had guarded as a treasure: incomparably accurate observations of the course of Mars across the sky.

The orbit of Mars was the key to the description of the heavens. It was irregular and capricious, impossible to fit into any astronomical system. Brahe's heirs certainly did not intend to allow Kepler free access to this painstaking work. But Mars' orbit was something he just *had* to find out about – so he simply stole the notes.

Using Brahe's decades of observations, Johann Kepler correctly described the solar system based on Copernicanism. With pen and paper he calculated, using a series of observations, the movements of a planet – Mars – that had a certain orbit. These observations were not, of course, made from a "fixed point" – that was what made them difficult – but from another mobile planet, the Earth, that was moving on an entirely *different* orbit. He had to do this without knowing the exact shape of the orbits in advance, far less their circumferences.

He published his calculations in the book *The New Astronomy*, which came out in 1609. In it he demonstrated that the orbit of Mars with its seemingly unexpected capers across the sky could be explained simply and correctly from two fundamental assumptions. One was Copernican: that the Sun stood still and was orbited by Mars and the Earth.

The other assumption was, in a way, a yet more radical break with the entirety of Aristotelian thought. For no one – and certainly not Copernicus – had been able to conceive that the planets' orbits would be anything but circular. The circle was the perfect shape, the classical symbol of perfection, where every point was equidistant from the centre.

But Kepler did his calculations. And he discovered that the planetary orbits were not divinely perfect circles, but earthly, bulging ellipses, figures that do not even have a centre, just two "foci", of which the Sun was one. He also demonstrated a peculiar proportionality, "Kepler's Second Law", that should have gladdened Galileo's heart: the area an imaginary line from the Sun to a planet "sweeps across", is always proportional to the planet's periodic time, regardless of how the distance between that planet and the Sun varies.

This was serious stuff from Kepler. He was not talking about mathematical models, but giving a factual description of cosmological reality, a description that also had the merit of providing correct calculations.

There was nothing he could do to solve the parallax problem. That aside, he had, if not proved, at least shown it to be overwhelmingly likely that Copernicus was right. It must be stated, though, that this was not the main point for Kepler. It was the great astronomical revelations of a religious, metaphysical nature that he really wanted to find and describe.

The educated world reacted with dismay, wonder – and temporary silence. Galileo said nothing on this occasion either.

The following year, in March 1610, one of the Emperor's most senior counsellors arrived at Kepler's house in his carriage, he was excited and bore extraordinary tidings. There were rumours at court that a mathematician in Padua had looked at the sky through a telescope and seen four new planets!

Kepler waited expectantly for further details. There was little point in looking himself as the telescopes available in Prague could barely be used to make out large and undetailed characteristics on the surface of the Moon. But he did not need to wait long. A few days later he had post from Galileo, the first contact between them for thirteen years. It was *The Starry Message*.

Even before he received the book, Kepler realised that what the Italian had seen must be satellites. He was not slow in working out that this was a weighty argument in favour of Copernicus. Strictly speaking, the existence of Jupiter's satellites proved nothing about what sort of centre the *planets* revolved around. But it was a serious warning that one of the basic assumptions of Aristotle and Ptolemy was crumbling. Satellites orbiting Jupiter would demonstrate that the Earth was not the centre of *all* cosmic motion.

According to Kepler's logic, the circumstantial Copernican evidence was proof enough that the satellites existed. In considerable haste he sat down and wrote a glowing defence of Galileo and the heavenly bodies he himself had never seen. The work was full of digressions, some brilliantly incisive, others comparatively speculative. Amongst the latter must be included a brief account of the kind of building styles hypothetical Moon-dwellers might employ. He sent his work by the first courier to Italy, but retained a copy that he polished up a little more and had printed.

He was able to send the printed piece to Magini, when the professor wrote from Bologna trying to mobilise Kepler against Galileo. His covering letter was dry and formal:

> "Accept this and excuse me. Both of us [Galileo and himself] are Copernicans. Like sees like company (...)."[21]

Enthusiastic support from Kepler, the Imperial Mathematician and Astronomer, was just what Galileo needed. Perhaps it was meant more as a lifeline to Copernicanism rather than to Galileo personally, but it arrived at precisely the right moment. Grand Duke Cosimo could now be certain that the Medicean stars really were up there around Jupiter to the eternal glory of his family.

Galileo never returned the favour. He did not even reply to Kepler's letter or thank him for the favourable wind generated by the printed pamphlet. In August he received yet another letter in which Kepler said he had received several other applications from Italy, where the satellites were still in doubt. Kepler distanced himself vehemently from Martin Horky's libel, but added that Galileo must get independent verification of his observations as soon as possible.

Now Galileo *had* to answer and quickly, too. Kepler's support was too vital to be hazarded. It was an amiable, but empty letter in which Galileo carefully avoided promising to send Kepler a telescope – although that summer he sent telescopes to prominent people all over Europe via the Medici's Ambassadorial network. (It was one of these that Kepler finally managed to borrow.)

We do not know for certain whether Galileo had read *The New Astronomy*, but even before 1612 at latest, he was well acquainted with the insights it contained. He never used them, however. Perhaps it was Kepler's bombastic style, full of digressions and odd assertions, that scared him. Galileo's *The Starry Message* is as different from Kepler's writings as it is possible to get; a concise, crystal clear elucidation, in fact the beginnings of the modern scientific style.

Or perhaps it was pure envy.

When Emperor Rudolf died in 1612, Johann Kepler wisely retired from the uneasy court at Prague where the whole of Europe's national and religious disparities were clashing. He settled in the provincial town of Linz. When his mother was implicated in a witch trial, he had to begin a legal and theological battle to save her. But he still found time to work. His last great book *World Harmony* (1619), was full of mystical speculations, but also contained his "third law": the square of a planet's periodic time is proportional to the cube of its orbit's greatest radius.

However, Kepler did not regard this law as a fundamental key to understanding the mechanics of planet movement – Newton, over fifty years later, was the first to do that. Instead, Kepler believed he had here found a proof for his planet mysticism, the divine harmony that must suffuse the world.

This work made no impression on Galileo either. The Grand Duke's Mathematician in Tuscany never allowed himself to be persuaded by the Imperial one in Austria. (Kepler kept his title for the remainder of his life.)

It is a strange historical paradox that it was the superstitious Kepler, with at least one foot in the mysticism of the Middle Ages, who proved the correctness of the Copernican system. The sober Galileo carried on the self-assured Tuscan tradition of sceptical, independent thought. A good century before, Leonardo da Vinci had announced that astrology was an untenable science whose main function was to get money out of fools. It might just as well have been said by Galileo (which did not prevent him from occasionally casting horoscopes in the course of his duties or for the sake of amusement).

Galileo was in many ways a modern rationalist. Even so, he did not manage to *prove* that Copernicus was right – and worse still, he repudiated or ignored Kepler's proofs. If, despite all this, he was impressed and influenced by his colleague's deep insights, he never admitted it to anyone. Perhaps the proud Tuscan viewed Kepler not as a brilliant astronomer, a collaborator in the work of gaining new knowledge, but as his major rival in being the principal harbinger of the Copernican truths in Europe.

Several Signs in the Sky

After more than twenty years Galileo returned home to Florence. Both his mother and the elder of his two sisters, Virginia, lived in the city, and for the first few months Galileo rented his living accommodation from his sister and brother-in-law, until he could organise a home of his own.

But it was not only a homecoming, it was also a leave-taking. He left behind Sarpi, Sagredo and other good friends in Padua and Venice, as well as an employer – the Venetian Senate – which was not pleased at his departure, as he had accepted the agreement of employment for life. Some of his friends thought he was acting very rashly in leaving the relatively liberal Venice. In Tuscany his room for manoeuvre would be dependent on the favour of the Grand Duke.

He was also leaving Marina Gamba after more than ten years together. The "family" was split up. His son Vincenzio, who was only four, remained in Padua. His two daughters went with Galileo to Florence – in fact, Virginia, his elder daughter, was there already. She had accompanied Galileo's mother home to Florence after one of the latter's visits. Patently pleased at having saved her grandchild from Marina's clutches, old Giulia wrote: "The girl is so happy here that she won't hear the other place mentioned any more."

What actually happened to Marina, is not known. With her modest background she clearly did not belong at the side of a mathematician to a grand duke, and not, perhaps, in the rather haughty city of Florence at all. Many of Galileo's biographers say that she married in Padua and that Galileo sent her money for Vincenzio's keep, but this is hardly likely. It is more probable that she died a short time after Galileo left, and that his son was fostered by a couple whom Galileo knew[22].

Young Grand Duke Cosimo ruled over a city and a grand duchy that was losing more and more of its relevance in European finance, culture and politics. Florence was no longer an international centre of power for trade and banking, but more a self-satisfied provincial capital in a rich, traditionally agricultural area. The great artistic challenges were to be found in Rome, where papal commissions for works of art large and small was the driving force. Galileo's friend, Cigoli, had moved there. But the truth was that, after two centuries of Italian dominance, the most exciting things were happening elsewhere: painting was flourishing in Rubens' Holland, literature in Shakespeare's England. Only in music had Italy retained its leading position.

Cosimo's political ambitions were not lacking. He wanted nothing less than to organise a new crusade, and liberate the Holy Sepulchre from Turkish rule. In the realms of reality he got no further than fetching a dubious Arab chieftain to Florence, a man who claimed to be able to raise a revolt amongst local tribes who were dissatisfied with the Turkish administration. He and his retinue strutted around the city at Cosimo's expense arousing much interest. Cosimo II was certainly no crusader, he was sickly and often ill; in addition he was strongly influenced by his wife and mother – who, in turn, did not always agree amongst themselves.

However, the life of the court was as magnificent as before. The aristocracy of Florence and other guests constantly had to be amused and entertained. Galileo was to help by adding lustre to the court, but his job was actually rather nebulous. He had no definite duties and he never visited Pisa where he was nominally employed. He was not of noble birth, and therefore could not be included in the innermost circles of court life. On the other hand, his telescopic observations had brought him international renown.

The house he bought had a roof patio where he could set up his telescope. He made more and better instruments and had them beautifully finished – one was covered in leather with gilded decoration, not unlike the binding of a book.

His fame did not exactly wane during the autumn of 1610. He received a secret communication from the French court earnestly entreating him to discover other new bodies in the heavens, so that King Henri IV could also be represented in the firmament. This never materialised, but to make up for it, independent observations of Jupiter's satellites began to pour in.

Galileo did make two new discoveries. To protect his priority, and also prevent others from hearing about what he had found before he himself had verified and publicised his findings, he coded the discoveries in anagrams which he sent to reliable people like Christopher Clavius at the Collegio

Justus Sustermans' painting of Galileo in the 1630s is the classical portrait: an alert, vital and yet venerable man. (© Photographic Archive, Institute and Museum of the History of Science – Florence)

A self-assured absolutist ruler: Duke Cosimo I as portrayed by his court painter Agnolo Bronzino. (© Corbis)

Villa Medici. The Tuscan Embassy and Galileo's regular quarters in Rome. (© Corbis)

The death of Giordano Bruno. Relief from the Bruno Monument on Campo dei Fiori in Rome (Ettore Ferrari, 1889). (© akg-images)

Johann Kepler. Mathematician to the Imperial court, visionary, dreamer and genius. (© Photographic Archive, Institute and Museum of the History of Science – Florence)

"The most brilliant and learned doctor, Nicolas Copernicus... Incomparable astronomer." (© Photographic Archive, Institute and Museum of the History of Science – Florence)

Robert Bellarmine. Teacher, theologian and cardinal – but above all an enemy of heresy in all its forms. (© Photographic Archive, Institute and Museum of the History of Science – Florence)

Maffeo Barberini as the representative of Rome's liberal intelligentsia – portrait from 1601, by Caravaggio. (From Riccardo Bassani e Fiora Bellini's: *Caravaggio Assassino*, Rome 1994)

Urban VIII as the incarnation of papal authority – bust by Bernini, 1637. (© Corbis)

Romano and Johann Kepler. The latter was very inquisitive and tried persistently, but unsuccessfully, to break the code.

The first discovery concerned the planet Saturn, the outermost of the planets then known. It lay on the very limit of what Galileo's best telescope could discern, and what he saw he described as two small satellites close in to the planet. However, they vanished from his observations before he could name them. These were really the rings of Saturn, but it was to be another fifty years before they were correctly described.

The other one was more important. He "publicised" the find in the mysterious anagram *Haec immatura a me iam frustra leguntur o y*, which means something like "These immature things are brought together in vain by me".

This was a reference to Venus. If Copernicus was right, and Venus orbited the Sun, the planet should display "phases" in the same way as the Moon. When the planet is furthest away from the Earth, it is fully illuminated by the Sun and thus "full". Gradually, as it proceeds on its course, the part illuminated by the Sun makes up less and less of what can be observed, and when it is between the Sun and the Earth, the lit half is turned away from us, and Venus is to all intents and purposes invisible.

But Venus was hard to observe; it was so bright that it caused colour refraction in the primitive lenses. In late autumn 1610 the planet was well positioned in the evening sky, and Galileo had an improved telescope to hand. For three months he observed Venus carefully, and he was left in no doubt. By December he could give the answer to his anagram: *Cynthiae figuras aemulatur mater amorum* – "The mother of love [Venus] imitates the appearances of Cynthia [the Moon]".

After only one year with telescopic observations, the Copernicans had strengthened their case considerably. Galileo was in the process of coming out openly with his view. The international fame gave him a platform, his discoveries spoke for themselves. Even in *The Starry Message* there was a cautious passage on how the existence of Jupiter's satellites undermined the theory of the Earth being the definitive centre of the universe. In letters and conversations he went a lot further.

The Church's experts had to assess the stream of new observations quickly. And the Church's foremost astronomers were Galileo's old friends the learned Jesuits of the Collegio Romano. Their leader was still Christopher Clavius, now an old man, but as immersed in his subject as ever. To begin with he was highly sceptical. Cigoli wrote in October: "Clavius said to one of my friends about the four stars that he laughs at them."[23]

Galileo took the initiative on behalf of the Medicean stars. He sent letters to Clavius, and he invited the Jesuits in Florence to look through his own telescope. They came and were convinced – according to Galileo's own report.

But the need for any influence was unnecessary. Father Clavius was a thoroughly honest scientist. All he needed was a better telescope. Once he had assured himself of the correctness of Galileo's observations, he at once wrote a respectful letter to the Grand Duke's mathematician congratulating him on his pioneering work: "Truly Your Lordship deserves much praise since you are the first to have observed this." Clavius put in some observations he himself had made, and urged Galileo to continue his work: perhaps he would discover "other new things about the other planets".[24]

Naturally the Jesuits understood as well as all other astronomers that the Ptolemaic system would be untenable if one accepted Galileo's observations of the night sky. But this did not mean they could accept the ideas of Copernicus, which contradicted the direct word of Scripture. Instead they eventually settled on a hybrid model, launched by the great Tycho Brahe, and which he believed was his great and lasting contribution to astronomy.

This *Tychonic* view of the world assumed that the Earth stands still, that the Sun orbits the Earth and that the planets, in their turn, orbit the Sun. This was both theologically acceptable and in keeping with the observations that had been made up to then – and, furthermore, it resolved the problems connected with birds' flight and falling bodies, which many people believed were unavoidable if one assumed that the Earth revolved on its own axis. These were objections that Galileo had largely dealt with at Padua, but had not publicised.

He now wanted to go to Rome himself to discuss his ideas and discoveries with leading colleagues. But he fell ill almost as soon as he arrived in Florence; a recurring illness with aches and fever which was to plague him for the rest of his life. But the letter from Clavius pepped him up. He replied to the old Jesuit immediately:

> "[Your Reverence's letter] has in large measure lifted me from my illness, since it has brought me the acquisition of so good a witness to the truth of my observations."

Galileo went on to bemoan the most sceptical of his opponents:

> "They are waiting for me to find a way to make at least one of four Medicean planets come down from heaven to earth to certify that they exist..."[25]

Despite encouraging reactions, the inland winter of Florence was quite a change from the mild, damp ones by the Adriatic that he had now become acclimatised to. His illness lingered on. It was not until March 1611 that he could set off on his trip to Rome. Grand Duke Cosimo encouraged him in this enterprise and told his Ambassador at the papal court to give Galileo all necessary support. They had to make the most of the family's triumph in the skies.

Friendship and Power

Just one year earlier Galileo had been a respected professor in the small city of Padua, unknown outside academic circles. Now he came to Rome, the capital of the world – *caput mundi* – only to discover that he was a celebrity whom people almost fell over themselves to invite and cultivate. His old friend and well-wisher, Cardinal del Monte, wrote a letter to Cosimo:

> "If we were still living under the old Republic of Rome, I verily believe that they would have erected a statue in the Capitol to honour his superb skill."[26]

Galileo lived in style at the Tuscan Ambassador's Villa Medici. The day after he had arrived, he visited the Jesuits at the Collegio Romano. He was heartily received. After discussing the telescopic observations thoroughly, Clavius and his colleagues decided to invite interested Romans to a public lecture, with Galileo in attendance. There they would explain his discoveries, and the Jesuit astronomers could announce that they too had made observations that supported Galileo's.

This address, about and for Galileo, was just as much a social event as a scientific lecture. The entire upper echelon of influential Romans, both within and without the Church, seated themselves in the Jesuit college's great hall. There they learnt about the irregular surface of the moon, Jupiter's sensational satellites, and of the new, extraordinary phenomena that had been observed in conjunction with Saturn and Venus.

One of the most enthusiastic members of the audience was a compatriot of Galileo's, four years his junior, an erudite jurist who was the scion of a well-known Florentine family, but who had lost his father at a young age and had grown up with an uncle in Rome. There he received a wonderful education at the hands of the Jesuits and went on to study law.

Maffeo Barberini was interested in all things new. He used his legal training to carve out an ecclesiastical career for himself, but he was not overly keen on dogma and theological niceties. Instead he cultivated art, literature and science in a private academy and in the circle around del Monte. Here they discussed painting, played music and did chemical experiments, but it was poetry that he found most engaging. He wrote himself, in a polished style and in Latin.

Barberini was a gifted man. Pope Clement VIII noticed him early and selected him for an honourable and delicate task.

Much papal cunning and calculation was needed to navigate the waters between the two Catholic powers, conservative Spain and the more liberal France. In 1601 Henri IV – a much talked-of convert from Protestantism – and his Maria de' Medici had had their first son. The papal court naturally had to be represented at the baptism of the heir to the throne, and the Pope selected the young Florentine Barberini.

Maffeo Barberini celebrated this leap forward in his career in a remarkable way. He commissioned a portrait from the most radical and controversial painter in Rome, Caravaggio. The portrait shows a self-assured and eager, but also sensitive young man, firmly grasping the commission assigning him his noble task.

Barberini's visit to Paris was a huge success. With his open and intelligent manner he charmed both the king and queen. In 1604 he returned to the French court as papal *nuntius*, a position equivalent to that of Ambassador. Relations between Paris and the Vatican were strained, partly because Henri IV had banned the Jesuit order from France, and would only allow it back in under strict conditions. But the sovereigns got on excellently with Maffeo Barberini personally.

In 1605 Paul V Borghese became the new pope. He, too, thought much of Barberini's efforts. When the Florentine returned to Rome, he was elevated to cardinal. Now, at the age of just 38 he found himself on the penultimate rung of the Catholic Church's ladder.

Maffeo Barberini was fascinated by what was happening on the frontiers – and not just in the world of painting. He was equally interested in the discoveries that had been made using Galileo's telescope. Dogmatics and mathematics he left to the theological and scientific pedants respectively.

Barberini left the lecture in the Collegio Romano ecstatic and enthusiastic. His degree in law was from Pisa, and he possibly had a superficial acquaintanceship with Galileo from there. Now he made contact. The two men had much in common, both in age and background. The cardinal and the mathematician hit it off immediately and became friends.

Another notable theologian was more troubled by Galileo's discoveries. Cardinal Robert Bellarmine, the formidable antagonist of Bruno and Sarpi, was not content to judge Galileo's assertions from just one passing lecture. For safety's sake he wrote a letter to his Jesuit brothers at the Collegio Romano asking whether all the new things he had heard were right. Clavius and the other astronomers could only acknowledge that Galileo was correct.

Ludovico Cigoli, who had known Galileo from his youth, also wanted to pay his own tribute. The painter ensured that he honoured his friend in a very special way. He was in the middle of a highly prestigious commission, helping to decorate a side chapel – the Cappella Borghese – of Santa Maria Maggiore, one of Rome's most important churches. The commission had come from the Pope himself, who intended to use it as his own chapel of rest when the time came.

Cigoli was painting a Virgin Mary on the chapel ceiling, in which she is standing on a moon. This moon is painted exactly according to Galileo's observations, with rings of mountain ranges and irregularities, not as the perfect, Aristotelian sphere.

Cigoli had no imitators in this experiment – and he was certainly in dangerous theological waters, no matter how correct he was on the astronomical side. A moon with scars and blotches was hardly a suitable symbol for *Maria Immacolata*, the pure and unadulterated Virgin!

However, the most important mark of esteem Galileo received during his triumphal progress through Rome did not come from ecclesiastical or artistic quarters, but from a representative of the most elevated Roman aristocracy.

On 14 April 1611 Galileo was invited to dinner at a fashionable villa on the Gianicolo hill. He took along his telescope so that the other guests might try it. These included a Greek mathematician, Demisiani, who coined a fitting name for the new instrument, made up of the Greek words for "distant" and "see": *telescopum*.

The dinner had been arranged so that Galileo could meet an extraordinary young man. Prince Frederico Cesi was only 28 years old. His parents had tried to smother his interest in science by sending him away from Rome, but it was hopeless. When his father died and Cesi came into both title and money, he founded his own independent scientific academy at the age of twenty, intended to promote natural research free from the shackles of academic tradition and the Church's scepticism towards everything new.

Prince Cesi was rich, but like other members of the old Roman aristocracy, he was gradually being ruined by the costs of maintaining a high lifestyle, while families with new names like Aldobrandini, Barberini, Borghese and Chigi had scions elected as popes and thus were financially and socially upwardly mobile. The parvenu-ridden social life that could be enjoyed at the papal court had no appeal for Cesi the aristocrat, even though he often had to attend. Instead he would go out to his country estates and observe nature. These observations were then discussed at his *Accademia dei Lincei* – "Academy of the Lynxes".

The name had been chosen because the lynx is commonly credited with having unusually sharp eyesight. Observation and perception, not accepted ideas, were to be the watchwords of their work.

The meeting between the Roman aristocrat and the Florentine scientist was a happy one, personally as well as scientifically. The pair became immediate friends, despite their disparate social backgrounds and the 18-year difference in age. Cesi needed Galileo's fame to lend his academy prestige, whereas the mathematician wanted solid contacts in Rome. Not least he wanted help in editing and printing the many books he planned to write.

A few days later, Galileo was admitted to the *Accademia dei Lincei* as its sixth member. Here, more than at the Grand Duke's court or amongst the intellectuals of Florence, he discovered his future scientific milieu, as his eager correspondence with Cesi and other members testifies. The social kudos that was to be gained from a connection with Prince Cesi, Duke of Aquasparta, Marquis of Monticelli, was not unwelcome either.

Paul V was not particularly interested in astronomy, but he, too, wanted to meet the man the whole city was talking about. Court mathematician and novice academy member, Galileo was invited to a formal audience.

Pope Paul well knew of the misgivings of his best theologian, Bellarmine. And he was certainly aware that Galileo had been a friend of the infamous Paolo Sarpi. But the Prince of the Church was at his most gracious during the meeting with this scientist who had brought such lustre and fame to the Italian states. As a special concession Galileo was not required to kneel for the entire conversation, as would normally have been the case.

A Dispute About Objects that Float in Water

A fortnight into June Galileo returned to Florence. Only a few days later he got embroiled, during the worst of the Tuscan summer heat, in a discussion about ice.

Galileo was staying with his good friend, the wealthy Salviati, at the Villa delle Selve – "the house by the woods". It was beautifully situated on a prominence near the little town of Signa, roughly half way between Florence and Pisa. From the noble house with its simple, firm Renaissance style, orchards of olive trees and vineyards stretched all the way down to the Arno. Salviati would retire here with his friends when the summer heat became too oppressive in Florence.

Filippo Salviati was deeply interested in science. Visiting his villa at the same time were two professors from Pisa, and for some reason they began to discuss the nature of ice.

Aristotle said that when things cooled they condensed. Clearly, ice was cooled water – and thus condensed, according to the two professors. Ergo ice was *heavier* than water.

But – Galileo objected – ice actually *floated* in water. If one believed Archimedes, would this not mean it was *lighter*?

Nonsense, the pair persisted. Heaviness and lightness had nothing to do with the characteristics of floating, because Aristotle had never said anything about it. (It was true that the master had hardly touched on the matter – he had in all only written one and a half pages on bodies that float and sink.) *Shape* was what was decisive. Ice floats on water because it is flat and thin.

Galileo knew he was right. In principle shape was irrelevant – it was specific gravity that determined if something floated or not, and he expressed himself in no uncertain terms to the two professors.

Perhaps the little argument in Salviati's villa might have ended there. But a few days later the disconcerted Aristotelians from Pisa were encouraged in the strongest terms to defend themselves and their scientific tradition against the arrogant Galileo. The previously mentioned local philosopher, Ludovico delle Colombe, had joined the contretemps.

About the same age as Galileo, delle Colombe clearly had some old grudge against him. He had written a short discourse on the 1604 nova, which had been torn to shreds by a certain "Alimberto Mauri", and he believed that Galileo was behind it. He had read Horky's attack on the telescope with great glee.

And delle Colombe had followed Horky's example and attacked Galileo in writing. But he went one decisive step further and shifted the battle about science and prestige into a new arena: that of theology. Without actually naming Galileo, he tried to hit him in the most dangerous place: where astronomy and Bible study converge.

In *Against the Motion of the Earth* delle Colombe wrote:

> "Could these poor fellows [namely, the promoters of the Copernican theory] perhaps have recourse to an interpretation of the Scripture different than the literal sense? Definitely not, because all theologians, without exception, say that when Scripture can be understood literally, it ought never be interpreted differently."[27]

Earlier that summer, when Galileo had still been in Rome, delle Colombe had tried to enlist several like-minded people in a campaign against the Grand Duke's mathematician. He wrote to Christopher Clavius and complained of Galileo's observations of the Moon's uneven surface and the implications this might have. His aim was clearly to provoke the Jesuit Clavius into a theological and scientific discussion with Galileo, but Clavius had seen the Moon for himself and was convinced. He did not even bother to reply to the Florentine philosopher.

Ludovico delle Colombe now heard about the ice argument. He immediately contacted the professors at Pisa and not only told them that, of course, Aristotle and they were right – but more to the point, he could prove that Galileo was mistaken. And he could defeat this jumped-up grand ducal Mathematician on his home ground – he could prove it with the aid of an experiment!

What ensued was as much a battle for status as for scientific truth. Cigoli saw it plainly when he wrote from Rome, making a reference to delle Colombe's odd surname:

> "Those ugly birds want to make a name for themselves not through their own value but through the choice of adversary."[28]

What delle Colombe was doing, was in effect to challenge Galileo to a duel. Not with actual weapons, but through the agency of a public experiment that would prove which of them was right.

The experiment was remarkably simple and anyone could understand it. No one doubted that ivory had a greater specific gravity than water. Normally, therefore, a piece of ivory would sink. But if one took a small splinter of the material and placed it carefully on the surface – look for yourself! It floats! Ergo there was the proof that shape *had* an effect on the ability to float.

Neither Galileo nor anyone else had the least inkling that this concerned a phenomenon called surface tension, and therefore had little to do with the general ability to float. It seemed indisputable that delle Colombe had a good argument.

The discussion therefore became one about how the experiment was to be conducted. Either Galileo wanted the splinter to be wetted before it was placed on the surface of the water, or he wanted to turn the experiment round so that it would demonstrate what *floated up* from the bottom of a vessel – a splinter of ivory, no matter how small, would, of course, stay at he bottom. Galileo's strength lay in the fact that he could think up lots of experiments that would prove him right, whereas his opponent was dependent on his single one, which was simple and appeared convincing.

The result was just about a draw. The Grand Duke was not pleased that his mathematician was in public dispute with delle Colombe, the more so since the latter had found support from a somewhat dubious Medici relative, Cosimo I's illegitimate son Giovanni. Galileo therefore withdrew from the entire proceeding, and instead wrote a dissertation in which he interpreted the whole problem in his own way: *Discorso alle cose che stanno in su l'acqua o che in quella si muovono* or *Discourse on floating bodies in water or those which move in it.*

Of its title there is this to say: firstly, this dissertation was written in Italian, not Latin. From now on it was the well informed general public that Galileo was addressing. His writings were to be accessible and comprehensible to courtiers and the bourgeoisie, and not just to scientists – indeed, he was excluding foreign colleagues like Kepler, who normally could not read Italian, or at least would have to struggle through the text with the help of their knowledge of Latin.

Secondly, typically enough he does not give way one iota to delle Colombe. Galileo had to be right – on all points.

He will explain how something can both float on water, and sink. In his dissertation Galileo attempts an astute explanation for the uncomfortable fact that tiny pieces of heavy materials do actually float. He refers to an instance he regards as analogous, namely that an empty clay pot floats, even though fired clay is heavier than water. But one must also include the volume within the pot as well, or as he puts it "the sum of the air and the material".

This – which is obviously quite correct – he transfers to floating fragments. He believes that they sink a tiny bit below the surface, without breaking it, thus forming an "air pocket" above them. The volume of this air must be included, and that is how the ivory floats.

This reasoning was brilliantly conceived – but erroneous. It cannot be denied that, on his own premises, delle Colombe was right in a way. And in addition to his cock-suredness, it was Galileo's innate tendency to search for the simplest and most rational explanations that caused problems for him.

He could not accept that the surface of water could have properties in any way different to water in general.

The Aristotelians were not to be convinced, and they soon launched a counter-attack.

Regardless of this, *Discourse on floating bodies* was a marvellous work, that linked the general comments Galileo had made on motion, to an investigation of things that move through water. In particular he emphasised that there is no "lightness" that can lift objects – in opposition to a "heaviness" that makes them fall. This is a fundamental assumption for Aristotle, linked to his doctrine of the elements: fire and air move upwards, water and earth downwards. In wishing to get rid of "lightness", he was aiming at the very foundation of Aristotelian physics.

Before *Discourse on floating bodies* came out, Galileo took part in a debate. It was not a public affair in the city, but rather on his home ground, at the Grand Duke's court. His opponent was no delle Colombe of modest social status, either, but the Aristotelian Papazzoni, the newly appointed professor at Pisa. The discussion was pure show, an intellectual entertainment Cosimo II had arranged for two very eminent guests after a splendid banquet.

Both guests were cardinals, and they threw themselves wholeheartedly into the discussion, fired just a little, perhaps, by the food and certainly the wine. Cardinal Gonzaga sided with Papazzoni, while the other supported Galileo, who indubitably came out of the skirmish best.

This other cardinal was Galileo's friend and admirer from his stay in Rome, Maffeo Barberini, whose star was continually on the ascendant in the clerical firmament. He was fascinated by Galileo's intellectual audacity, and was not too worried if Aristotelian doctrine was brought to its knees. Such a supporter in the Vatican could be a good person to have, because now his enemies were rallying for the attack.

Sun, Stand Thou Still upon Gibeon!

Galileo had imagined a productive existence in his home city – with no teaching, and circumstances easy enough to preclude the need to manufacture instruments or rent out rooms to make ends meet. He had plans for several books, and of course he would continue his observations with the telescope.

He also continued to work on his Copernican ideas. Prince Cesi proved himself a perceptive correspondent, well versed in Kepler's new ideas. In a letter from the summer of 1612 he discusses if any of the heavenly bodies might move round the Earth or the Sun without these necessarily being the exact centre of the orbits. He adds: "… and perhaps everything moves in this way, if the planets' orbits are elliptical, as Kepler has it."

But Galileo did not manage to accomplish everything he wanted. He was often ill, with severe bouts of fever every year or second year. His symptoms sound rather like an hereditary, periodic fever ("Mediterranean fever"), which also causes joint pains resembling rheumatism, something Galileo was very plagued by. The type of rheumatism he suffered from aside, it was not helped by his fondness for wine, which raises the uric acid levels in the blood.

He was also responsible for two young daughters. They were quite unalike. Virginia, the elder, was extrovert and lively, intelligent and – to judge from her later letters – "daddy's girl". Her sister Livia on the other hand, displayed a tendency to melancholy – it was likely she had closer ties to her mother and missed her more.

Galileo had planned similar futures for both of them – he wished to put them into a convent as soon as possible. Respectable marriages were out of the question, as they had been born "out of wedlock". He attempted to mobilise his old benefactor, Cardinal del Monte in Rome, to try to get a dispensation from the rule that nuns must be at least sixteen before taking their conventual vows, but it was no good. In the meantime, in 1613, he boarded them at the convent of San Matteo at Arcetri just outside Florence when they were thirteen and twelve years old respectively. They would live there until they could become nuns.

But also the controversies surrounding his person and his ideas began to take up a lot of his time and energy. In December 1611, Cigoli wrote from Rome:

> "I have heard (…) that a gathering of ill-disposed men who are jealous of your talents and your fame, met at the house of the Archbishop [of Florence] and put their heads together…"

He also hints that there are supposed to be plans to ask a priest to declare from the pulpit that Galileo "says extravagant things"[29].

Philosophers who whittled ivory splinters to defend their Aristotelian positions were one thing. Priests and archbishops, quite another.

But he did manage to do a little astronomical observation during this period. A small, almost forgotten sighting shows how Galileo with his practical sense had developed into a phenomenally skilful practitioner with the telescope in the space of a couple of years. He could genuinely, and quite literally, congratulate himself on the "lynx-like vision" Prince Cesi believed all the members of his academy ought to possess.

Around New Year 1613, he glimpsed with his still primitive telescope an unknown, dimly glowing body in the vicinity of Jupiter. He noted the find, but the object vanished after a few days, and he did not follow it up. Everything points to the fact that Galileo had caught a glimpse of the as yet unknown planet Neptune, which was not discovered and described until 1846, more than two hundred years later.

Of more immediate importance was another heavenly body. Pointing a telescope directly at the Sun was not particularly wise. But Galileo learnt to project the light from the Sun, through the telescope and on to a sheet of paper. There he could study the Sun's disc minutely. The most striking things were the dark, mobile areas that appeared on it. He called them *macchie solari* – "sunspots".

Pretty quickly Galileo discovered that these sunspots gave two further arguments against traditional cosmology. In the first place it was clear that the Sun was no more perfect or immutable than the Moon. Secondly, the movement of the sunspots strongly suggested that the Sun rotated on its own axis – in just the same way as Copernicus' opponents said it was impossible for the Earth to do.

So, still no proof, but more and more circumstantial evidence.

Germany possessed an able Jesuit astronomer by the name of Father Christopher Scheiner, who was also interested in sunspots. With their well developed international web of contacts, the Jesuits had got hold of good telescopes, and Scheiner was a fine observer. He now wrote a short account in which he began a discussion with Galileo about sunspots. One of his assertions was that he had seen the phenomenon before the Italian.

But Galileo regarded telescopic observations as his own private domain. His reply to Scheiner – or *Apelles*, as Scheiner had called himself – was published by the Accademia dei Lincei under the title *Letters on Sunspots* with a preface that could be read as patronising.

Old Father Clavius of the Collegio Romano died in 1612, and his successors in Rome were not, perhaps, on quite such good terms with Galileo. Although their discussions were couched in the politest terms and expressed mutual respect, a certain reserve began to insinuate itself between the influential Jesuits and Prince Cesi's Academy of the Lynxes. In reality Cesi did nothing to lessen the clash, on the contrary, he regarded his academy as an alternative to the religiously dominated scientific institutions. In fact he had expressly enacted that monks and priests were barred from membership.

This Father Christopher Scheiner would prove to be a man with a very long memory, and he was just as touchy as Galileo himself. But for the time being he was polite and reserved, as befitted a Jesuit and a scientist. He replied to Galileo from a more principled starting point. Scheiner wrote (using the name of a pupil) a short book with a long title, *Mathematical Discourses on Astronomical Controversies and New Discoveries*, in which he argued against Copernicus both from the mathematical and biblical standpoints. He sent the pamphlet to Galileo and evidently hoped for a cultivated discussion between scientists who disagreed purely on professional matters.

Discourse on floating bodies also caused controversy. No less than four books were published that set themselves against Galileo's ideas. One of them was written by the indefatigable Ludovico delle Colombe, who by then had begun to call himself an "anti-Galilean". These opponents were of such humble social status (and scientific status for that matter) that it would have been unseemly for the Grand Duke's mathematician to answer them. According to the custom of the times, Galileo let his best student from Padua, Father Benedetto Castelli, who had become Professor of Mathematics at Pisa, do the answering. This humiliation did not make "the league of doves", as the group eventually became known, any the less intractable.

An odd episode occurred in November 1612, which at first made Galileo furious, but which he subsequently joked about. An elderly Dominican priest in Florence, named Lorini, said during a discussion that as far as he could make out, maintaining that the Earth moved was contrary to Holy Scripture. When Galileo wrote and demanded an explanation, the Dominican replied, apparently defensively, that his remark had been made off the cuff, that he did not have the slightest knowledge of astronomy, or at least not of that "Ipernicus or whatever his name is".

Galileo laughed at the naive priest – triumphantly and far, far too soon. For whether this was pure chance or planned – the next cut was aimed quite differently.

A year later Professor Castelli was lunching with the Grand Duke, who was then staying at his palace in Pisa. The talk centred around Galileo, the telescope and astronomy in general, and another professor who was present said that, for his own part, he was definitely of the opinion that the theory that the Earth moved, was contrary to the teaching of the Bible.

Also present at the lunch was a devout and earnest woman who held the Bible's word in deepest respect: Dowager Grand Duchess Christina, Cosimo's influential mother. Even though Castelli made light of the episode and believed that he had stopped his colleague's mouth, Galileo was worried.

Key to the discussion around the Grand Duke's lunch table was an Old Testament passage from the Book of Joshua, chapter 10, verses 12–13. It deals with a settling of scores between the Israelites and one of their warlike neighbouring tribes, in this instance the Amorites. The Lord waded in with a terrible hailstorm ("... they were more which died with hailstones than they whom the children of Israel slew with the sword"). But Israel's general, Joshua, was not content. He needed more time to complete the massacre of the enemy, and so he offered up this prayer: "Sun, stand thou still upon Gibeon; and thou, Moon, in the valley of Ajalon." God hears his prayer (v. 13): "And the sun stood still, and the moon stayed, until the people had avenged themselves upon their enemies. (...) So the sun stood still in the midst of heaven, and hasted not to go down about a whole day."

The next verse underlines the miraculous nature of the event: "And there was no day like that before it or after it (...)."

If the Lord could, by a miracle, make the Sun stand still, the implication had to be that it normally moved. Therefore there was an open conflict between Scripture's unambiguous words and the Copernican theory. To dismantle the entire ingenious Aristotelian-Ptolemaic philosophical edifice would have profound consequences. It would alter enlightened laymen's picture of the world and undermine the prestige of traditional academics. But Galileo knew only too well that those few words in the Book of Joshua had far more weight for many of his opponents. Scriptural interpretation was not an area for private discussion. In the spirit of the Counter-Reformation, everything of this sort was the absolute monopoly of the Church.

The Letter to Castelli

Despite incipient opposition from ecclesiastical quarters, Galileo nevertheless felt fairly secure. The Pope had received him well. The Jesuit astronomers were on his side. The antagonism caused by the discussion about sunspots, was still barely noticeable. The powerful Robert Bellarmine was sceptical it was true, but to counterbalance that, Galileo had a friend and admirer in Maffeo Barberini, who was close to the Pope.

The Cardinal and the mathematician still kept in contact. Galileo wrote and told of his discoveries. Maffeo Barberini replied in cordial tones. When Galileo became ill in October 1611, the Cardinal wrote at once assuring him of his devotion, wishing him a speedy recovery and hoping that, for the benefit of all, Galileo would enjoy a long life. Uniquely amongst Galileo's many correspondents, he did not end his letters with the ornate and somewhat high-flown assurances of esteem common at the time. Maffeo Barberini signed quite simply *come fratello* – "like a brother".

And then, Galileo was under the protection of the Grand Duke of Tuscany, even though Cosimo was constantly ill and had handed over more and more to the family's two strong women, his mother and his wife.

Taken as a whole, the Grand Duke's mathematician thought that any misunderstandings between cosmology and biblical passages, ought to be clarifiable. And he saw no reason why he should not do this himself. For the first time this took him into the realm of theology. In a long letter to his former pupil and constant friend, the Benedictine monk and professor, Father Castelli, he sketched out his fundamental views on the relationship between Holy Writ and natural forces.

Galileo saw both as manifestations of the divine, and as such there could never be any real disagreement between them. These apparent collisions arose because Scripture has to be tailored to human understanding and comprehension. Forcefully he enumerated the misinterpretations that could come about if one was not clear on this point:

> "... grave heresy and blasphemy, for in that case it will seem needful to give the Lord hands and feet and eyes, and also carnal and human feelings, such as anger, remorse and hate, and sometimes He will have forgotten the past and be ignorant of the future."[30]

The language of the Bible had to be interpreted. If a natural phenomenon showed itself to be inescapably true, the theologians would have to go back to the Bible, see how that truth was formulated within it, and elucidate the relevant passages in the light of nature's revelation of the divine.

Concerning the secondary conclusions the Church drew from various biblical passages, Galileo thought – and this perhaps not without a trace of irony – that it would be safest not to postulate more articles of faith and dogma than those absolutely essential for belief and salvation.

He ended by giving his own reading of the passage from the Book of Joshua. Very much true to character, he did not select the simplest and most obvious interpretation: that "And the sun stood still" was merely a figurative way of speaking, an illustration of God's miraculous ability to suspend the forces of nature. On the contrary, he used a subtle line of argument to try to show that the passage could be read as a justification for Copernicus!

Castelli thought the letter convincing – so convincing that he had it copied and used it in further discussions about the motion of the Earth, the

Bible's words and the Church's authority. For the discussion continued in Florence, amongst clerics and in other public forums. But neither Galileo nor Castelli took much notice of all the talk, until the matter suddenly blew up a few days before Christmas 1614, a year after the "Letter to Castelli" had been written.

Precisely what Cigoli had hinted at three years previously, occurred. Father Tommaso Caccini mounted the pulpit of the Dominicans' mother-church, Santa Maria Novella, and delivered a blazing sermon that began with a text from the Acts of the Apostles 1,11: "Ye men of Galilee, why stand ye gazing up into heaven?" The play on words was especially effective in Latin: "Viri Galilaei" could just as easily mean "Galileo's men" as "men of Galilee".

Caccini took the well known verses from the Book of Joshua and gave them a thorough and literal interpretation. After which he attacked all those who believed differently, in other words Copernicus and his adherents. In a rousing conclusion he pronounced mathematics to be one of the Devil's many arts, and that mathematicians should be driven out of all the Italian states because they spread false teaching.

The attack had come from Caccini personally, or to be more precise, from a cleric who represented the anti-Galilean faction in Florence. It had not been cleared with the Dominican authorities or the Vatican. Even so, Galileo was livid. Letters of support from other Dominicans, who distanced themselves from the views of their Florentine colleague, gave him some succour. But not enough. Galileo wrote to Prince Cesi in Rome to discuss what ought to be done. Even though the attack had been directed at him, the consequences of it affected the study of a mathematically based description of the world.

Cesi's reply was a cold shower. That worldly Roman replied that the whole affair had to be handled with the greatest delicacy. The reason was Robert Bellarmine.

This ageing, but still powerful Jesuit had personally told Prince Cesi that he considered Copernicus' doctrine to be heretical, and that he believed that the notion of the Earth moving was unquestionably at variance with the words of Holy Scripture. Bellarmine, who was a hugely learned man and had studied the physics and geography of the time intensively, assumed that Dante's poetical image of the world represented reality: Hell was quite literally at the centre of the Earth, Heaven was the "outermost" sphere in a closed universe[31].

Galileo was not happy with Cesi's answer. But he let the matter rest, all the same. His local antagonists had not finished with him, however. Now, old Father Lorini, with his "Ipernicus or whatever his name is", came back on to the scene.

Lorini had got hold of a copy of the "Letter to Castelli". He read it, and found its contents required him to take the matter up with members of his order at the monastery of San Marco. Everyone agreed that it was a very serious matter.

The Dominicans – commonly known as "God's dogs", *Domini canes* – were, together with the Jesuits, the Church's front-line soldiers in the war against heresy. The formal leadership of the Inquisition rested with them. Just like the Jesuits, the order placed great emphasis on scholarship, but it was more orientated towards philosophy and theology than natural science. The Dominicans were highly sceptical of the rapid growth of the Jesuit order. There was much rivalry between the two organisations, although few would have gone so far as the Dominican who announced that he crossed himself each time he met a Jesuit! The fact that Galileo was still, apparently, in favour with the Jesuits, would not necessarily make him less suspect in the eyes of the Dominicans – possibly quite the opposite.

When old Father Lorini read out the "Letter to Castelli" to his brothers, he set a lengthy process in train. God's guard dogs in San Marco scented heresy, and they gave their warning loudly and clearly.

It was no longer just the Copernican system and its dubious relationship to the unambiguous words of Scripture that was the problem. In his attempt to reconcile faith and science, Galileo had transgressed another boundary, one that had been clearly drawn up by the ideologues of the Counter-Reformation during the Council of Trent:

> "... the Council declares that in matters of faith and morals pertaining to the edification of Christian doctrine, no one relying on his own judgment and distorting the Sacred Scriptures according to his own conception shall dare to interpret them contrary to that sense which Holy Mother Church, to whom it belongs to judge of their true sense and meaning..."[32]

In short, this might indicate that Galileo had himself committed what was the Lutherans' and the other reformed churches' gravest sin: begun to interpret the Bible for himself.

For Father Lorini there was only one thing left to do: he had to report the matter to the Vatican. On 7 February 1615 he sent a copy of the letter to Rome. Its destination was the office with the momentous title *Congregation of the Index of Prohibited Books.*

"How to Go to Heaven, Not How the Heavens Go"

Now Galileo spied the danger. It was no longer a matter of a local quarrel in Florence, episodes the like of which he, with his position at the Grand Duke's court, could smile at condescendingly. The threat was so serious that he must meet it on two fronts, partly in Rome, partly at home in the court.

Things were not made better when Galileo learnt that Father Tommaso Caccini, the man who thought that mathematicians ought to be exiled, had gone to Rome. He was to take up a position at the important Dominican monastery in the church of Santa Maria sopra Minerva. Galileo anticipated that Caccini would use his new position to continue the attack on him.

Galileo realised that he had to mobilise factions in Rome that were kindly disposed to him personally and not over-sceptical towards Copernicus. This principally encompassed the Jesuit mathematicians. The professor who had taken over from Christopher Clavius at the Collegio Romano was called Father Grienberger. Through the offices of a friend, Galileo forwarded a copy of the "Letter to Castelli", begging that this – *correct* – version be given to Grienberger and then sent on to Bellarmine, "if an opportunity should present itself". Galileo added that Copernicus had been "not only Catholic, but religious and canonical".

"The correct version" were his own words. There was a number of minor discrepancies between the copy Father Lorini had sent, and the one Galileo now forwarded himself. The differences did not concern the fundamentals, but in Lorini's version they consistently showed Galileo in a worse light than in his own.

So strong has been the sympathy for Galileo in posterity that all his biographers have accepted the interpretation given by Favaro, the publisher of Galileo's collected works: Lorini's copy was purposely slanted against Galileo. The most recent research however indicates something different: that it was Galileo's new "copy" that was slightly toned down and edited in comparison with the original letter.

For unknown reasons, Father Lorini had forwarded the "Letter to Castelli" to the wrong institution. *The Congregation for the Index* worked in tandem with the Inquisition and its task was to produce a list of books Catholics were not allowed to read, *Index librorum prohibitorum*. But the letter had not been printed, and so did not come under the Congregation's jurisdiction. It was therefore passed over to the Inquisition. Here it was routinely read by a theologian-consultor, who quickly expressed an opinion. He pointed to three unfortunate formulations (all of which were different in

Galileo's new "correct" version), but concluded that the letter was nothing to get worried about.

The Inquisition's plenum of cardinals was not thoroughly satisfied however, and would not let the matter rest there.

In the meantime Galileo, through an intermediary, learnt of the reactions of the Jesuit Grienberger – and Cardinal Bellarmine. Neither was particularly positive. Bellarmine said straight out that Galileo should regard Copernicus' system purely as a mathematical model. In such a case its relationship to the words of Scripture would present no difficulty at all. In addition, he threw another biblical passage into the debate, the Book of Psalms, 19, 5–6: "[the Sun] Which is as a bridegroom coming out of his chamber, and rejoiceth as a strong man to run a race. His going forth is from the end of the heaven, and his circuit unto the ends of it (...)."

Grienberger said that he would prefer Galileo to provide clear observations before he was drawn into discussing Scripture. Apart from this his reaction was cautious, though not unfriendly, but it was quite obvious that Galileo could no longer depend on the warm support he had received from the Jesuits when Clavius was alive.

But Galileo ignored the opposition. He had taken new courage. He wrote a letter in which he clearly stated that Copernicus was serious and was not merely postulating a mathematical model. Either one accepted that the Earth moved and the Sun stood still, or one did not – but in the latter case one was making a grave error, something Galileo intended to demonstrate in a work he was engaged on. To reinforce the point he concluded the letter with a home-spun Copernican interpretation of the passage in the Book of Psalms, precisely the sort of activity which he, as a layman, ought to have kept well away from. His correspondent in Rome immediately went to Prince Cesi with the letter, and they quickly agreed *not* to show it to Bellarmine.

One reason for Galileo's defiance was that he had suddenly found support in an unlikely theological quarter. Father Foscarini, a Carmelite monk from Naples, made public a letter he had sent to the head of the Carmelite order, in which, with professional theological sophistry, he had argued in favour of Copernicus – and Galileo. He divided the problematic biblical passages into six classes and suggested six exegetic principles that would solve the problems.

This impressive construction did not help much. Bellarmine was also asked for his opinion of this work, and it was not high. Behind the series of courteous fraternal niceties that were expected between sons of the Church, his meaning was crystal clear: Copernicus' system could be used for purposes of calculation, but *definitely not* to explain reality. Certainly Bellarmine used all the subjunctive reservation of which the Italian language was capable, that *if* some day it could be proved irrefutably that Copernicus was right, then one would have to go back and interpret the relevant biblical passages again. But he excluded the possibility of such proof.[33] The doctrine that the Earth was in motion not only went against common sense, the Book of Joshua and the Psalms of David, but against Solomon himself; Solomon who had gained all his wisdom from God. For was not this ascribed to Solomon in the Book of Ecclesiastes: "The sun also ariseth, and the sun goeth down, and hasteth to his place where he arose." (Eccles. 1,5).

The other reason for Galileo's strong stand, was that he believed he had just the indisputable, physical proof that Bellarmine required.

But before he went public with his new proof he wanted to make certain of his home ground. Galileo decided to combine a theological and theoretical account with a courtly tribute and write an open letter to the Dowager Grand Duchess. The "Letter to Christina" ran to over forty pages, and circulated only in manuscript copies, as any attempt to publish it would have risked an open confrontation with the censor.

In the letter he makes his position clear. Truth is one and indivisible. There can therefore be no conflict between the words of the Bible and natural revelations, but the Bible is written in a different language and has a different object: it teaches us "how to go to heaven, and not how the heavens go." (*Non come va il cielo, ma come si va in cielo.*) This implies that the Bible's words must be explained and interpreted.

After this, Galileo goes on the offensive. He tries to enlist one of the Church fathers on his side. Bellarmine had repeatedly pointed out that the entire theological tradition was against Copernicus' ideas. But Galileo takes the case of Augustine, and believes he can show that he has an anticipatory and wholly different position regarding questions of natural science. In conclusion, he returns to his clarification and introduces an enlarged and overhauled Copernican version of the Sun miracle in the Book of Joshua.

The "Letter to Christina" was principally written for a "home readership", to create back-up from the Grand Duke's family. Presumably it succeeded in this. But as far as Rome was concerned it did little to help.

There things were happening fast. The motive force was still the Dominicans from Florence. In his new role at Santa Maria sopra Minerva, Father Caccini now had direct access to the heads of the order. He contacted one of the Holy Office's top men saying that "for the sake of his conscience" he wanted to make a declaration about Galileo's errant ways.

His deposition was a mixture of fact, rumour and insinuation. Caccini correctly pointed out that Galileo thought that the Earth revolved on its own axis and orbited a static Sun. He went on to say – as everyone present naturally knew, as they had heard the account of the "Letter to Castelli" – that the mathematician had embarked on the dangerous practice of producing his own interpretation of Scripture.

As these well known pieces of information obviously were insufficient, Caccini went a step further. He claimed that another Dominican in Florence had heard the disrespectful way some of Galileo's followers had spoken of God and his saints. In addition, he aired Galileo's old friendship with the infamous Paolo Sarpi in Venice, and opined that the two still corresponded by letter. (This was correct – Galileo wrote to the ageing monk, telling him about his discoveries.) By way of conclusion he emphasised the dubious aspects of the Accademia dei Lincei – especially that its academicians patently corresponded with Germans.

Everyone knew there were Lutherans in Germany.

The heads of the Inquisition decided that the matter had to be looked into more carefully. As usual they did their work thoroughly, and used most of 1615 in getting to the bottom of the matter. The Inquisitor in Florence carried out interrogations and Galileo's works and letters were carefully read and commented on. This was all supposed to happen in the greatest secrecy, but it was impossible for Galileo not to know that something was afoot.

He *knew* that Copernicus was right. If the Catholic Church was definitely going to put the whole of its power behind the opposite view, the consequences would not only be a terrible reverse for scientific study in Italy, but the Lutherans in the north would triumph, and attract men of talent because of the relative freedom of ideas that existed there.

Galileo recalled his triumphal progress in Rome four years previously. Now he was ill and unable to work for long periods. Even so, he believed it was imperative for him to return in person to bolster his friends and win round doubters and opponents. He must make the Jesuit astronomers show their true colours, and ensure Maffeo Barberini's continued friendship and support. He must argue objectively with the sceptical, but highly intelligent Bellarmine, and get him to see through the untenable arguments of men of the "league of doves" calibre.

If possible, he must get another audience with Pope Paul V.

For he had his new, incontrovertible argument up his sleeve. It was complicated, but if necessary he must try to lay it directly before the Holy Father.

Foolish and Absurd in Philosophy, Formally Heretical

Galileo's supporters in Rome were not looking forward to his visit. They feared that his eagerness and conviction would only make things worse. Far better for him to stay at home, working away quietly at his arguments.

The Tuscan Ambassador who would have to be his host, wrote to the Grand Duke's Secretary of State:

> "... this is no fit place to argue about the Moon or, especially in these times, to try and bring in new ideas."[34]

The Ambassador was quite right, but it made no difference. With gracious permission from Grand Duke Cosimo, Galileo arrived in Rome in December 1615.

To begin with he behaved as if he was still the feted and celebrated observer of Jupiter from four years before. His cock-suredness completely prevented him from absorbing the scepticism and repugnance with which he was met in many quarters. On the contrary, he was in better spirits than he had been for a long time. His visits to leading Romans took the form of lectures and dazzling discussions, as if he was still amongst his admiring students from Padua.

If, during the conversation in these elegant salons, a doubting cleric or nobleman objected that the Earth could not revolve in just one day, that such speed was unthinkable, Galileo would turn the argument around and point out that, according to Ptolemy, the entire *constellation* revolved in one day, and that was unimaginably larger than this planet. If they took up the old argument that the Earth's motion must at least be noticeable by us, Galileo would invite them to think that they were aboard a ship: let a ball sink slowly in a container of water while the ship is at rest. It will sink straight down, without touching the sides. But if the ship is under way at a constant speed – what happens to the ball then? It still sinks straight down. It is not affected by the even motion of the ship.

These were impressive intellectual demonstrations. But they did not help the matter in hand. People who got thoroughly trounced in such discussions hardly looked on Galileo with any greater good will. Slowly this became apparent to him.

It was time to play his trump card, his irrefutable new argument. In January 1616 he sent a letter to one of his adherents in the College of Cardinals, the very young Alessandro Orsini. The letter was a treatise on the tides, the causes of high and low water.

In the days when he had frequently travelled between Padua and Venice, Galileo had noticed the great barges that carried fresh water across the lagoon and into the city. The water was contained in large, open vessels, and when the barges changed speed for any reason, the water moved. If the speed was reduced, the water sloshed forward, rising at the front end of the container and sinking at the back.

As usual, Galileo was amazingly quick to make a connection between a physical observation and an underlying principle: the sea was like the water in the container, and the boat was the Earth. In this way, high and low tide could easily be explained, *but only if one assumed that the Earth moved!*

Working out a theoretical justification, did indeed prove difficult, not least because, in discussion, he had so splendidly demonstrated that the Earth's motion did *not* influence other movements, an absolutely central plank in the argument against opponents of Copernicus. But the tides *had* to be explained that way, if they were to constitute a directly observable proof that the Earth moved.

He ended up with a complicated reasoning that took into account both the annual movement of the Earth round the Sun and the daily revolution of the Earth round its axis. The theory did have to have several additions bolted on to it, taking account of sea depths, narrow inlets and the like, to explain the large local variations in tides, but these did not shake Galileo's conviction. If one could only follow his reasoning one could, right then and there, and with one's own eyes, see clear proof of Copernican ideas – just the kind of proof that Bellarmine had predicted would never be found.

His theory of tidal movements – which Galileo doggedly clung to until well into his old age – was, however, completely wrong. But *that* was not the deciding factor in the events that now followed in quick succession.

The Holy Office had at length come to the conclusion that Father Caccini's string of allegations against Galileo were too diffuse. The matter had to be shelved. But at the same time many churchmen looked on with disquiet and distaste at the dazzling way the mathematician carried on his Copernican propaganda right in the heart of Rome, the bastion of Christendom.

And this caused Galileo's "salvaging expedition" to set in train just what he wanted to avoid. The cardinals decided to attack the issue from another direction. It was unnecessary to hit Galileo directly. The ideas of *Copernicus* were the problem. If they were prohibited, all discussion would cease.

Now things moved at lightning speed. No lengthy investigations were needed, for Copernican ideas were well known. The Inquisition's leaders had a meeting and formulated two assertions which the cardinals believed summed up the Copernican view. They then handed them over to a group of experts for an evaluation and conclusion.

The assertions were these:

> "That the Sun is the centre of the world and hence immovable of local motion."
>
> "That the Earth is not the centre of the world, nor immovable, but moves according to the whole of itself, also with a diurnal motion."[35]

The Inquisition's experts took four days over their work. The majority were Dominicans with only one Jesuit. Their expertise lay solely in the theological field, none of them had any qualifications in astronomy. What happened was exactly what Galileo had most feared: Copernicus' theories were condemned from a literal reading of Holy Scripture, without assessing one single material physical or astronomical argument. The conclusion, which the cardinals of the Holy Office unanimously adopted, was this:

The first assertion was

> "foolish and absurd in philosophy, and formally heretical, since it explicitly contradicts in many places the sense of Holy Scripture (...)"[36]

The other assertion was given

> "the same censure [qualification] in philosophy and that in regard to theological truth it is at least erroneous in faith."[37]

That the assertion concerning the Sun as the world's immobile centre was "formally [*formaliter*] heretical", did not simply mean that it was a formal mistake to postulate it. On the contrary, the wording implied the grossest possible censure. Anyone who, in the future, maintained that the Sun stood still, would be viewed as a pure heretic – and would have to accept the consequences of it.

This resolution was passed by the Inquisition on 24 February 1616. That same day, Cardinal Orsini tried to advance Galileo's tidal theory to Paul V. The timing was the worst imaginable, and the offensive failed completely. The Pope said that the best thing Orsini could do was to rid Galileo of his delusions. When the Cardinal continued to argue, the Pope cut him off abruptly. No sooner had Orsini left the room, than the Pope summoned another cardinal, one with very different views: Robert Bellarmine.

The Hammer of the Heretics

The Jesuit and cardinal, Robert Bellarmine had dedicated his life to the fight against heresy in all its forms. Physically, he was a small figure. But this did not prevent him from radiating personal authority and he could make the most powerful of men shrink when he gave them one of his penetrating stares. His admiring co-religionists called him "the hammer of the heretics". On his grave in the Jesuit mother-church of Il Gesù was this telling inscription: "By force have I subdued the thoughts of the strong."[38] Over three centuries later he was canonised as San Roberto after one of the most controversial processes in the Church's history.

The problem of Galileo had not finally been laid to rest by the Inquisition's resolution. But now there were formal grounds for halting his crusade in the name of Copernicus, but it ought to be done quickly – and preferably as discreetly as possible. From a political point of view it would be unfortunate if Galileo lost face and was publicly humiliated. Grand Duke Cosimo was hardly likely to take it kindly, and that could mean quite an unnecessary worsening in the relations between the Papal States and the Grand Duchy of Tuscany.

But Bellarmine had ways and means. He suggested a plan of action to the Pope that was sanctioned by a plenum of the Inquisition the following day. Galileo would be given a clear, but private, warning. If he refused to take notice of it, the warning would be made official in the name of the Inquisition. In the unlikely event that this did no good either, the mathematician would be put in gaol, whether he was the Grand Duke's man or no.

The next day Galileo was called in to Bellarmine's official residence, the Paradise Rooms in the Vatican Palace. Segizzi, another cardinal from the Inquisition was also present.

Bellarmine was the one who did the talking. He used his authority to issue a warning, couched in firm but friendly phrases: the decision of the Inquisition had to be respected. It entailed that the Copernican system should not be portrayed as a factual picture of physical reality. Under no circumstances was it permitted to maintain that the Sun actually stood still, or that the Earth moved round it.

But Galileo was not a man to give in just like that, not even to the combined authority of Bellarmine and the Inquisition. He *had* to protest. Not only was the prohibition a personal injustice and a terrible setback for the work he had done over the past few years, it was monumentally stupid and erroneous. He had the proof himself, the tides! It was impossible to hold back: Galileo began to argue with Bellarmine.

At this, Cardinal Segizzi cut in. A friendly warning about who was making the decisions was obviously not getting through. Presumably, he thought Bellarmine's attitude was far too mild. With the entire weight of the Inquisition behind him, he gave Galileo orders not to teach, defend or discuss the prohibited assertions.

This ended the meeting. But the powerful Bellarmine was offended by Segizzi's brusque intervention. He felt quite able to manage the affair according to his own plan, especially as he had the Pope's express authority to proceed cautiously. So he refused to sign the account of the meeting that Segizzi's notary had prepared. At the next plenary session of the Holy Office he gave a brief resumé of how he had settled the matter using his own subtle diplomacy.

Bewildered and broken, Galileo returned to his lodgings in the Villa Medici after his confrontation with the two Cardinals. Still, he was not completely crushed. As he had understood Bellarmine, the warning only applied to direct propaganda in favour of Copernicus. Therefore it should still be possible to work quietly on the thing and, above all, use the heliocentric system as a mathematical hypothesis, something indeed that Bellarmine had always been prepared to accept. Bellarmine had smoothed over Cardinal Segizzi's outburst.

But Segizzi's account of the meeting found its way into the archives of the Holy Office.[39]

The Inquisition could only do half the work of removing the Copernican aberration. The rest was up to the Congregation of the Index. Bellarmine also had a place on that. As early as 5 March there was a public decree temporarily prohibiting Copernicus' *De Revolutionibus Orbium Coelestium* – a book that had been legal reading for seventy years – in anticipation of necessary alterations. Father Foscarini's new book, however, with its theological defence of Copernicus, was totally banned: it was "altogether prohibited and condemned".[40]

In the course of just a fortnight, everything the Tuscan Ambassador had feared had come to pass. Galileo's burning enthusiasm had led to the exact opposite of what he had hoped for when he set out from Florence. Now the Ambassador begged the Grand Duke and his Secretary of State to recall Galileo as soon as possible, before something even worse happened.

For Galileo, who in the course of that harrowing February had celebrated his fifty-second birthday, was certainly not a broken man. Copernicus' book had not been prohibited for ever. It had to be "corrected", something which could surely be effected with an assurance that it described a hypothetical model and not the physical reality. It was also reassuring that he was called in to an audience with Pope Paul V just one week after the Index decree. The Pope's tone was amiable, he assured Galileo that the Church respected him both personally and as a scientist, and was disinclined to listen to gossip about him – provided he kept to the guidelines that had clearly been laid down.

Even so, rumours began to circulate in Rome. It was said that Galileo had been to see Bellarmine, where he had officially been required to renounce his Copernican beliefs after which he was given a heavy penance. Resolute as always in matters of honour and hearsay, Galileo requested a written denial of this, which Bellarmine willingly supplied. He wrote a letter in which he denied that there was ever any question of repudiation or penance[41]. The only thing that had happened was that he – Bellarmine – had informed Galileo of the decision reached by the Holy Office.

Segizzi's tactless intervention was not mentioned.

Galileo took the letter and travelled home to Florence, as he had by now received courteous, but firm orders to return. The Ambassador in Rome was more than happy to see the back of his troublesome guest:

> "... he is not at all in a good position for a place like this, and he might get himself and others into serious trouble."[42]

One name is largely missing from the accounts of Galileo's long stay in Rome in 1615–16, that of his friend and admirer, Cardinal Maffeo Barberini. He was an opponent of the interdicts but was powerless to do anything in the prevailing atmosphere. He had certainly become an influential cardinal but, against an alliance of the Pope and Bellarmine, he was impotent. Behind the scenes he worked to minimise the damage of the assault on the new cosmology. Maffeo Barberini was a member of the Congregation of the Index, and it was he who, together with a colleague, managed to avert a definitive ban on Copernicus' book.

One lone Italian stood up for Galileo, indeed Tommaso Campanella actually wrote a pamphlet *called* "In Defence of Galileo", *Apologia pro Galileo*. But this was assistance that Galileo could well have done without.

Campanella, like Giordano Bruno , was a Dominican from the Kingdom of Naples. He had studied at Padua for a year and knew Galileo from there. Then he was arrested and sent to Rome, just like Bruno, but released through the efforts of influential friends. He returned to southern Italy where, together with other Dominicans, he attempted to organise a veritable revolt against Spanish hegemony. The rebellion was easily quelled, and Campanella imprisoned in Naples where, oddly enough, he escaped the death sentence. He was eventually allowed a certain freedom to correspond from his cell, and he wrote repeated letters of admiration to Galileo:

> "All philosophers in the world now hang on your pen, for in truth one cannot philosophise without a true and certain system for how the planets are constructed."[43]

Galileo tried to keep his distance from this enthusiast, who almost seemed to be begging for friendship and scientific contact from his prison cell. Even though Campanella did not lack connections all the way up to the College of Cardinals, it was not enthusiastic support from a suspected heretic and convicted rebel that Galileo needed most. Campanella's defence was smuggled out of the Italian region and printed a few years later in Frankfurt, in 1622. No sooner had the first copies reached Rome, than the book was banned.

Campanella was himself no convinced Copernican. His defence – "an action that displays a most uncommon intellectual courage,"[44] as one Italian historian puts it – was ultimately a contribution in support of freedom of thought. He argued that the need to investigate how the world was created was a gift from God, and that it was therefore deeply un-Christian to place barriers in the way of such studies.

Deaths and Omens

Galileo's position had certainly been weakened, but not his self-confidence. He *knew* he was right and that the Holy Office, the Congregation of the Index and the Pope himself were wrong. Things could change. Society and Church politics in Rome shifted constantly, especially in conjunction with a change of popes. New men must necessarily one day come to the most important positions. Banned books could then be rehabilitated: it had happened before – even Bellarmine's first book was put on the *Index* because it was not sufficiently pope-friendly! At that time, though, Pope Sixtus V had died before the decision had been made public, and his successor had reversed it without delay.

Galileo's tactics can be seen in a carefully crafted letter he wrote to Grand Duchess Maria Maddalena's brother, Archduke Leopold of Austria. The letter accompanied a gift of two telescopes, his little book *Letters on Sunspots* and a handwritten copy of his reflections on the tides which he had sent to the young Cardinal Orsini. Of his analysis of the tides he writes:

> "With this I send you a treatise on the causes of the tides which I wrote at a time when the theologians were thinking of prohibiting Copernicus' book and the doctrine announced therein, which I then held to be true, until it pleased those gentlemen to prohibit the work and to declare the opinion to be false and contrary to Scripture. Now, knowing as I do that it behooves us to obey the decisions of the authorities and to believe them, since they are guided by a higher insight than any to which my humble mind can of itself attain, I consider this treatise which I send to you to be merely a poetical conceit, or a dream, and desire that Your Highness may take it as such (...). But even poets sometime attach a value to one or other of their fantasies, and I likewise attach a value to this fancy of mine. (...) I have also let a few exalted personages have copies, in order that in

> case anyone not belonging to the Church should try to appropriate my curious fancy, as has happened to me with many of my discoveries, these personages, being above all suspicion, may be able to bear witness that it was I who first dreamt of this chimera."[45]

Behind this letter lies not only Galileo's self-confident claim on priority as originator – of a "poetic fantasy"! – but also his deep and real fear that the lead in natural sciences would go to the Protestant and reformed countries of the north, where there were scientists "not belonging to the Church" and therefore not bound by the Inquisition and the *Index*.

At the same time he went on with his tireless work of turning his purely scientific discoveries into devices with practical applications – and thus, of course, into money. Galileo now had a brilliant solution to one of the greatest practical problems of his day: the fixing of longitude.

More and more international trade, not to mention international war, was conducted on the high seas. After the discovery of America and the sea route to India, great fleets of merchantmen and warships regularly sailed huge distances between continents – without ever knowing for certain where they were until they made land at some spot.

Latitude is a naturally determined phenomenon, defined by the poles and the equator. It can be fixed by the skilled observer by measuring the height of the sun or the angle between the horizon and a known star. Longitude, on the other hand, refers to a randomly selected starting point – the zero meridian – and has to be worked out with reference to that. The only practical method of doing this is to compare the local time at the ship's position with the time at the zero meridian. All one needs therefore is a totally accurate timepiece to take along on the voyage, giving standard time. Local time can be calculated from the sun's noonday height.

The problem was that clocks of this accuracy simply did not exist. The technology was not good enough, and furthermore, changes in temperature during the journey would affect the metal in all known alloys and cause inaccuracies in the clockwork. So, ships continued to run aground, seamen died of starvation and scurvy, precious cargoes were ruined, simply because it was impossible for the captain to know where on the map he had taken his crew.

Galileo was a landlubber who never stepped off the Italian peninsula. But he knew only too well that the great seafaring nations had promised a princely reward to anyone who could solve the problem of accurately fixing longitude. All that was needed was a precise clock. And he had one, a celestial one at that, visible to everybody: the satellites of Jupiter, or more accurately the eclipses of the four small moons. This lunar eclipse occurred roughly a thousand times a year – enough to enable the time to be taken, almost to the second, about three times a day.

Two things were required: extremely accurate tables for the thousand eclipses, and a piece of equipment that would enable a navigator on the ocean to observe the phenomenon with equal accuracy. Galileo set energetically to work producing both of these. He observed the satellites at every possible moment, and he constructed a huge instrument like a diving helmet with a telescope in front of one eye. He travelled in person to the Grand Duchy's principal port, Livorno, to test the equipment aboard a safely anchored ship.

But in practice the system was impracticable for shipping. In the first place Jupiter was not visible all year and never during the day – nor at night if the weather was overcast. Secondly, it was far too difficult to make such observations from a rolling ship's deck, even if the planet and its satellites were visible. It was to be more than a century before the English clockmaker, John Harrison, solved the longitude problem by constructing a clock that was accurate under all conditions.

If however, one was safely ashore and had time to wait a night or two for a good sighting, Galileo's method was excellent, and it assumed great importance for cartography in the second half of the 17th century.

Illness continued to plague the mathematician. In 1618 he tried a strange and uncharacteristic remedy: he set out on a pilgrimage.

There is no reason to doubt that Galileo, rationalist and sceptic though he was, reckoned himself anything other than a true Catholic believer. But this one, concrete manifestation of religious piety in him is still odd.

The object he chose for his journey lay in the small town of Loreto, on the Adriatic, scarcely 125 miles south-east of Florence. The way there was difficult, his route would cross the Apennines. Loreto boasted a reasonably sized church, a pilgrims' basilica that was built around a small wooden house ten metres by four. The house was the miraculous shrine that drew the pilgrims in.

The strangeness here lies in the fact that the miracle is connected with *motion*. Galileo chose to visit a relic that broke all the imaginable rules for the relocation of heavy bodies. The house in Loreto was supposed to be the *Santa casa* (the holy house), the Virgin Mary's home in Nazareth, where Jesus grew up. It had been borne through the air from the Holy Land by angels and set carefully down in Loreto in 1291.

It is extremely hard to conceive that Galileo believed in this miracle literally. In all the hundreds of pages he was to write on movement, there is not a single hint that natural laws can be suspended in this manner. As we know, he found the biblical miracles trying enough, without having to deal with all the ones supposed to have taken place in more recent times.

The trip to Loreto was no doubt partially occasioned by the house and its location being famed for miraculous cures. But it looks more like an attempt by Galileo to convince those around him – and perhaps himself as well – that his scientific work was occurring within the ecclesiastical framework that had been so strictly forced on him.

It might also have been sensible to trust to religion at a time when sudden death was a constant and unexpected visitor. In the summer of 1613 Galileo received a letter from Prince Cesi: his friend Cigoli was dead, barely 54 years old.[46] The considerate Cesi immediately visited his family to see if there was anything he could do for the bereaved on his own behalf, and on that of Galileo. The painter died suddenly, at the height of his career, shortly after Pope Paul had honoured him with the title of "Knight of Malta". He was in the middle of decorating the choir in one of Christendom's most important shrines, *San Paolo fuori le mura*, the church built over St. Paul's grave.

Cigoli would certainly have wanted to emulate Galileo and return home to Florence, but he never made it. Such reports of death were an everyday occurrence: rich Salviati with his retreat at Villa delle Selve died in 1614, Marina Gamba in 1611 or a little later, the wise and well-to-do Venetian Sagredo, arguably Galileo's best friend during the Padua years, in 1620.

These deaths affected Galileo's life in various ways. Now that he could no longer visit Salviati, he needed his own, spacious house in the hills, where the air was healthier and he could make his observations unhindered by lights or disturbance. In addition, he had to bring the last of his children to live with him, his eleven-year-old son, Vincenzio.

He rented an extensive house on the wooded ridge of Bellosguardo south-west of central Florence, not that far from the highest point of the enormous Boboli Gardens that Cosimo I had begun to construct above Palazzo Pitti. The villa was expensive – it cost a hundred scudi a year to rent – but Galileo reckoned that some of the expenses could be offset by the corn, beans, lentils and peas which he could produce on the large property.

He now also had the opportunity to grow grapes and make his own wine – a perfect combination of practical work and theoretical speculation. How did the juice of the sun-drenched grapes turn into alcoholic wine? His answer, which later generations of Florentine scientists pondered at length, was that "wine is a fusion of *umore* and light"[47]. *Umore* can mean juice, liquids generally, but is also the word for the four bodily fluids linked to different humours. (The role of yeast in the process was not discovered until the 19th century.)

He brought his son Vincenzio to the villa. And as if that were not enough, Galileo did what he had not done for his two girls: he ensured that his paternity was regularised by means of an official *leggitimazione*. His motivation was the same as his own father's had been when he had taken him away from the monks at Vallombrosa: Vincenzio was certainly not bound for a monastery, he was to be educated, earn money – and preferably get a decent dowry in marriage.

But Vincenzio was nothing like his loveable sibling Virginia, who had by now become Sister Maria Celeste in the nearby convent of San Matteo. Galileo soon discovered that it was no joke trying to be a father again at an age of well over fifty. He got little help from his own mother. A glimpse into relations with the aged widow Giulia is given by a remark in a letter to Galileo from his brother, the musician Michelangelo:

> "I hear, not with a little surprise, that Mother is again behaving so dreadfully. But she is very old now, so that will soon put an end to all this quarrelling."[48]

The quarrelling ended in 1620. His mother died in September at the age of eighty-two, after living for almost thirty years as a widow.

Comets Portend Disaster

Galileo's own health was not good. Fever and rheumaticky bouts kept him in bed for weeks on end. His work was set back a good deal. The worst thing was that his illness prevented him from observing the most interesting astronomical phenomenon of the period.

In fact it was a triple phenomenon, three comets of varying brightness that appeared in quick succession in the autumn of 1618.

These comets were the first to be visible in Europe after the invention of the telescope, and therefore of enormous astronomical interest. That they were also harbingers of disaster, as comets usually were, there could be little doubt: earlier that year rebellious Protestants in Prague had thrown three of the Emperor's henchmen out of the windows of the Hradcany Palace and had thus precipitated the pan-European catastrophe known subsequently as the Thirty Years War.

At all events these comets presaged a heated debate amongst astronomers and natural philosophers. There were two main trends of thought: those who still sided with Aristotle, who had of necessity to maintain that the comets were nearer than the Moon, they had to be part of the earthly and mutable. Aristotle assumed that comets were composed of vapours from the Earth which ignited when they ascended high enough, and were then driven by the Moon's sphere to move.

But Tycho Brahe had observed a comet in 1577 and, incomparably accurate observer that he was, had managed with the naked eye to take sufficiently good measurements to calculate the comet's parallax. From these measurements he concluded that comets had to be much further away than the Moon, probably somewhere near the orbit of Venus, and that they revolved around the Sun – not in circular, but in *oval* orbits.

But Galileo had not witnessed the phenomena himself. As a result he was at first little inclined to wade into the debate. But in 1619 a pamphlet entitled *An Astronomical Discussion of the Three Comets of 1618* appeared. The work was officially anonymous, but it was well known that the author was the man who now held the chair at the Collegio Romano that had once belonged to Clavius and Grienberger, the Jesuit priest, Orazio Grassi.

Father Grassi was a gifted, melancholic man. In an intelligent and well-meaning attempt to understand the comets he accepted a good deal of Tycho Brahe's argumentation. Grassi had access to Jesuit observations from all over Europe and could calculate the parallax. He admitted that the comets were further away than the Moon, and in so doing distanced himself from Aristotle. It was a step the Jesuits had already taken, as they had supported Galileo's telescopic observations. On the other hand, he could not follow Brahe in his assertion that comets moved round the Sun. As a material description of a heavenly phenomenon it was dangerously close to the prohibited Copernican ideas.

But in the discussion that raged in Rome's scientific and ecclesiastical circles, Grassi's assumptions were portrayed as a weighty argument *against* Copernicus. This irritated Galileo, but perhaps not so much as something else: Grassi's slender pamphlet does not mention Galileo once. In truth, the Grand Duke's mathematician had not distinguished himself in comet research. He had never personally observed a single comet, but it appears that Galileo continued to regard the astronomical use of the telescope as his own private and sacrosanct territory. Furthermore, he felt his position as Europe's most fashionable astronomer under threat. He was getting enquiries about the comets from several quarters including the French court. And he had nothing to say about them – whilst Grassi was bringing out new observations and theories.

He therefore made up his mind to reply. It is true that the pamphlet *Discourse on the Comets* was published in the name of one of his students, Mario Guiducci, but the surviving manuscript is almost entirely in Galileo's handwriting, and no one was in doubt as to its true author.

The pamphlet had a double objective. It was to repudiate *both* Aristotle's and Tycho Brahe 's comet theories. In so doing it would indirectly show that a new, third theory was required, one that, for very good reasons could not be postulated, because it had to be founded on Copernican thinking. But Galileo was too thorough in his repudiation. In order to best the Jesuit Grassi, he totally rejected Brahe's completely correct observations, and settled on the theory that comets actually *were* vapours from the Earth and they were closer than the Moon.

There was a "Copernican" basis for this faulty conclusion. If the comets and the Earth really did orbit the Sun, the comets should display *retrograde* movement in certain phases, in other words, move "backwards" through the sky during the periods when the Earth was "catching up" with them, just as the planets do. The fact that such movement cannot be observed, was used as an anti-Copernican argument. (The real reason is that comets are only visible from the Earth for a short period as they move towards the Sun.)

But the main motivation for Galileo's contention was probably psychological rather than astronomical. Grassi had promulgated an argument based on telescopic observations, namely that comets were not much enlarged by the instrument and therefore had to be correspondingly further away, an argument he believed not everyone had understood and accepted.

With his highly developed sense of pride Galileo managed to construe this as an attack upon himself – on the maestro of the telescope, the first, most skilful and practised telescope-user of them all! He felt he just *had* to counter such an argument. The result was that *Discourse on the Comets*, which was published in July 1619, not only built on a completely false concept of what comets were; but worse, the pamphlet was a stinging personal attack on Grassi and the entire Jesuit milieu around the Collegio Romano.

Honour was equally highly developed in that quarter. Orazio Grassi hit back the same autumn. Using the easily detectable pseudonym of Lotario Sarsi, he published *The Astronomical and Philosophical Balance*, in which proofs and assertions about comets were to be carefully weighed. The book is a strange mixture of stringent and up-dated thinking in the fields of astronomy and optics, mixed with uncritical quotations from the ancients – and some well-chosen, sarcastic attacks on Galileo:

> "I fancy I hear a small voice [in Galileo's text] whispering discreetly in my ear: the motion of the Earth. Get thee behind me thou evil word, offensive to truth and to pious ears! (...) But then certainly Galileo had no such idea, for I have never known him otherwise than pious and religious."[49]

At first Galileo did not realise how deeply he had offended Grassi, and refused to believe it was he who had written *The Balance*. But soon he had to recognise that it was indeed the case.

The truth was that in the course of the discussions about sunspots and comets, Galileo had managed thoroughly to irritate his most important scientific allies – the Jesuits in Rome.

But Galileo had been even more incautious. *Discourse on the Comets* had aimed some hefty swipes at Christopher Scheiner too, the German Jesuit astronomer that Galileo had fallen out with over the discovery of sunspots. Father Scheiner now lost all hope of any objective exchange of views with Galileo, which he had tried to encourage by sending him his *Mathematical Discourses* and a courteous accompanying letter some years before.

Grassi had to be answered. It was imperative for Galileo's honour and position, something to which his friends at the Accademia dei Lincei especially drew his attention. The affair also involved the academy's honour. A number of its members were highly critical of the Jesuits and were keen to see Grassi thoroughly taken down.

Nor was Galileo bereft of support in Rome – Maffeo Barberini continued to keep in contact. The Cardinal had great notions of his own worth as a poet, and he unleashed his talents in a private poetic eulogy to the mathematician – *Adulatio perniciosa*, in which he praised Galileo's discoveries of planets and sunspots. In the covering letter he wrote: "The respect I have always entertained for your person and for the virtues within you, have informed this composition which I enclose. I greet you with all my heart, in the hopes that Our Lord gives you contentment."

And, as he always did, Cardinal Barberini signed it – *come fratello*.

The shaping of his answer to Grassi took Galileo some time, however. Partly on account of caution, and partly illness. Unfortunately the pilgrimage to Loreto and the *Santa casa* had brought no relief.

In the meantime two more highly important people had died, Cardinal Bellarmine and the Pope himself, Paul V . The latter's successor was a typical compromise candidate, Gregory XV Ludovisi, an old cardinal from Bologna, who was remarkable for little except his incipient senility.

This papal election took place in February 1621, and it was clear that the new Pope would not increase the chances of freer expression on cosmological models, even though Bellarmine was gone. But almost at once a *third* important person died: Grand Duke Cosimo II. His son and heir, Ferdinando II, was only ten years old, so it was clear that his mother the Grand Duchess and his grandmother the Dowager Grand Duchess Christina would have an even tighter grip on the reins than before. What this would mean for the support Galileo might expect, was very unclear.

Galileo's response to Grassi was not ready until well into the autumn of 1622. By then it had, however, turned into an entire book which he sent to his Lyncean friends in Rome so that they could comment on the manuscript, get the censor's permission to print it (*imprimatur*), and carry out the printing process. All this took time, and Prince Cesi only had an edited manuscript ready for the press by the summer of 1623. Just then there was *another* in the long series of deaths that affected Galileo's destiny during these years. Pope Gregory XV passed away after two years on St Peter's throne.

The old Pope had managed to accomplish a couple of things. He had appointed Richelieu, the widowed French Queen Maria de' Medici's young counsellor, as a cardinal. And he had altered the rules governing the elections of future popes.

The new procedure meant that the election of Gregory's successor took a long time, for there was bitter conflict within the College of Cardinals. In the intense heat of the Roman summer the cardinals sat shut away in the Sistine Chapel for almost a month before they reached some kind of agreement. But on 6 August the white smoke drifted up from the chapel, and the College's spokesman emerged and cried out the longed-for words across Rome: *Habemus papam* – "We have a pope".

When Prince Cesi heard the news of who had finally won the papal nomination, he immediately halted the printing of Galileo's book. A new dedication was absolutely imperative. The book must be seen as a tribute to the new man in the Vatican, who had now taken the name Urban VIII.

This was Galileo's great opportunity. The new, absolute ruler of the Catholic Church's spiritual realm and the Papal States' temporal lands was the author of the eulogy *Adulatio perniciosa*, his admirer, countryman and friend "like a brother", the Florentine Maffeo Barberini.

Weighing the Words of Others on Gold Scales

> "I remain much obliged to Your Lordship for your continued affection towards me and mine and I wish to have the opportunity to do likewise to you assuring you that you will find in me a very ready disposition to

serve you out of respect for what you so merit and for the gratitude I owe you."[50]

Six weeks prior to the papal election, Cardinal Barberini, as he then was, had written these words to Galileo. Now he had been crowned with the tiara and dressed in papal vestments during a ceremony in which he had surprised everybody by prostrating himself on the floor of St Peter's before the altar and praying that God would end his life if his pontificate was anything but a blessing to the Church.

Galileo's own reaction to the news of the election of Maffeo Barberini can be judged from a congratulatory letter he sent to the new Pope's nephew, Francesco. Even allowing for the rhetorical exaggeration of the age, there is no doubt about the genuine enthusiasm:

"... how delightful it is for me to have whatever remains of my life, and how much less heavier than usual will death be at whatever moment it overtakes me: I will live most happy, the hope, up to now altogether buried, being revived to see the most unusual studies recalled from their long exile; and I will die content, having been alive at the most glorious success of the most loved and revered master that I had in the world, so that I would not be able to hope for nor desire other equal happiness."[51]

Galileo's re-kindled hope certainly was not unfounded. Young Francesco Barberini had just been made a member of the Accademia dei Lincei, and Urban VIII's first official action was to make his nephew a cardinal. Prince Cesi altered the rule about churchmen not being academicians for Francesco's sake. Other members of the Academy also stood high in the Pope's favour. One of these, Giovanni Ciampoli, was given the highly influential post of Papal Secretary and Privy Chamberlain. Suddenly Galileo had a number of contacts right inside the Church's hub of power.

But the appointment of his nephew Francesco was also the first sign of a new trait in Urban VIII Barberini. He certainly wanted to work for the glory of the Church, but at the same time he did a not insignificant job of concentrating wealth and office within his own family. The traces of this still survive in Rome today, and the family coat of arms with its three bees can be seen in many places. The principal monument is the generously proportioned Barberini Palace that contains Rome's National Gallery for older works or art.

As the years went by, the people of Rome, who were generally sceptical of papal power, became seriously annoyed. The worst thing of all was that Pope Urban removed the ancient bronze covering from the roof of the Pantheon, and recycled it on Bernini's lavish baroque baldachin over the Pope's altar in St. Peter's. As they said in the city: "What the Barbarians didn't do, the Barberinis did."

In the autumn Galileo's book came out, with its new dedication to Urban VIII, in which one of the more flowery passages went: "As we humbly bow down to your [Holiness'] feet, we pray you may continue to show favour to our studies with the well-disposed rays and strengthening warmth of your most goodly protection."

The text was in Italian, in contrast to Grassi's *Balance*, which was written in Latin. Galileo had hit upon a brilliant title, typical of his ever fertile talent as a polemicist: the book was called *Il saggiatore* – "The Assayer", the title of the office of the official controller of the purity of precious metals and the mixtures in alloys. The point being that assayers use weighing instruments which are much more accurate than normal, run-of-the-mill scales. They must use "gold-scales" for their careful calculations. Grassi's and Galileo's arguments were thus about to be given a *truly* accurate weighing!

The Assayer is a settling of scores with Grassi, alias Sarsi, regarding the nature of comets and their orbits. In this sense it is an unsuccessful work, as Galileo simply got the basic point about the essence of comets, wrong. He relied on his own ability to draw conclusions, and had never even observed comets properly through his telescope.

Comets did not fit into Galileo's ideal Copernican universe. They were unpredictable – and worse, if they really did move round the Sun, it was in those despicable elliptical orbits. Galileo did not want to hand any points to either Brahe or Kepler . He wanted to be the one to formulate the fundamental characteristics of the construction of the universe.

But such an evaluation of *The Assayer* is too simplistic. The discussion about comets – which is witty, acute and occasionally malicious – is merely the springboard for a general discussion of the potential for a description of nature, rich in examples and with innumerable openings for new challenges. And beneath it all lurks the forbidden faith in the Copernican system, from which Galileo distances himself with an irony so subtle that it is impossible to catch him out:

> "And since I could greatly fool myself in penetrating the true meaning of matters which by too great a margin go beyond the weakness of my brain, while leaving such determinations to the prudence of the masters of divinity, I will simply go on discussing these lower doctrines, declaring myself to be always prepared for every decree of superiors, despite whatever demonstration and experiment which would appear to be contrary."[52]

Grassi was made to feel like a cleric. Galileo carefully maintained the fiction that it was his pupil Mario Guiducci who had written *Discourse on the Comets*, and poured scorn on anyone who could think otherwise, people like "Lottario Sarsi, a completely unknown person" – the mis-spelling of the first name is perhaps a conscious word-play on the verb "lottare": to fight, wrestle.

Grassi had explained the title of his book *The Astronomical and Philosophical Balance* by saying that it was a reference to the constellation Libra, from which he believed one of the comets had appeared. Galileo claimed it was more likely to have come from Scorpio, and that Grassi's book was therefore an "astronomical and philosophical scorpion", which aimed a whole barrage of stings at him.

"But,"

he says,

> "it is my good fortune that I know the antidote and the remedies at hand for such stings! I will therefore break and rub that very scorpion on the wounds, where the poison reabsorbed by its own dead body will leave me free and healthy."[53]

Galileo maintains that he has retired from the public gaze because all his writings have been attacked and misunderstood, while others take the credit for all his discoveries; he is being "slandered, robbed and scorned", and his writings are refuted with "laughable and impossible notions". He expends a lot of righteous energy in demonstrating this.

After which he systematically tears "Sarsi's" text apart. As a polemicist Galileo stops at nothing. He consistently pretends that he has no inkling that Grassi is the real author of *The Balance*, and writes:

> "[Sarsi] repeats certain things he claims to have understood from Father Orazio Grassi, his teacher, concerning my latest findings; I believe not a word of this, and am certain that this priest has never either said, thought or seen Sarsi write such fantasies, they are so far removed from any respect for the doctrines by which teaching is done at the Collegio, where Father Grassi is a professor (...)."[54]

Obviously the Jesuits reacted to such statements. No reader could however fail to notice that between the shafts of sarcasm the sickness-wracked Galileo, almost sixty-years-old, shone with original observations, acute inferences and thought-provoking questions. It was in this book that he formulated his belief in mathematics – or rather geometry – as *the language of nature*. And he knew himself that this did not only apply to the regularities of cosmology, because he had the as yet unpublished revolutionary pendulum and free fall experiments from Padua up his sleeve, which he continued to develop during these years.

Related to his belief in geometry was his clear distinction between the fundamental and incidental characteristics of an object. The fundamental ones were exactly those that could be dealt with geometrically: shape, size, position, movement. But an object also has other traits which are interesting in themselves: colour, taste, smell. This latter group is different however because, according to Galileo, such characteristics are dependent upon someone sensing the object, and they can therefore be seen as fortuitous designations we associate with an object, merely "names" or "labels".

These speculations are leading towards the rudiments of an atomic theory. As with so many of Galileo's other ideas, this one was not new either. The notion that matter is built up of tiny, indivisible entities, goes back to Democritus in 400 B.C. But Galileo brings the idea in from the cold, and discovers that these phenomena, which we can so easily see and perceive in daily life, must be explained by means of something we *cannot* directly perceive or see. As regards light, there must be an "expansion and diffusion, rendering it capable of occupying immense spaces but its – I know not whether to say its subtlety, its rarity, its immateriality, or some other property which differs from all this and is nameless."[55]

Galileo wanted to get inside the phenomena, because they simply cannot be explained from our immediate sensual impressions. The idea of characteristics and atoms did not perhaps seem so obviously dangerous as his cosmological ideas had shown themselves to be. But Father Grassi, who of course was on the look out for exposed points for his counter-attack, noticed them.

In the first instance there was only one reader of *The Assayer* who really counted, and he was the one to whom the work had been hastily dedicated. Pope Urban VIII Barberini liked the book. He had nothing against sarcasm.

At least, as long as it was not directed at him.

It is possible that Galileo's Roman friends in the circle around the Accademia dei Lincei exaggerated the Pope's enthusiasm when they related that he had had *The Assayer* read aloud to him at mealtimes. But one of them probably did overstate things when he emphasised that *now* was the moment for Galileo to write down

> "those concepts which up until now remain in your mind, I am sure that they would be received most gratefully by Our Lord [the Pope], who does not cease to admire your eminence in all matters and to retain intact for you that affection which he has had for you in times gone by."[56]

Certainly Urban remained a firm friend and admirer of Galileo's, even after *The Assayer*. He was himself interested and intelligent enough to value its scientific speculations, and the wordsmith in him had to admire Galileo's pungent wit. In the unspoken, but important, competition between the Accademia dei Lincei and the Collegio Romano for the position of Rome's leading institution in the scientific sphere, the Pope almost had to be counted as one of the Accademia's supporters.

But he was not a Copernican.

There was one particular passage in *The Assayer* that Urban VIII admired, both for its linguistic elegance and its content. Galileo had introduced an extraordinary fable about the inquiry into sources of sound.[57] This told of an inquisitive man who, to his astonishment, discovered that similar sounds can have different origins: birdsong, the music of a flute, a bow drawn across violin strings, a man running his finger round the moistened edge of a glass. Finally, he finds a cicada, cannot work out how it makes its noise, examines it and at length discovers that it has some powerful chords under its thoracic shield. He decides to cut through these chords – if the cicada's song then stops, he will have found the source of the sound. But in his attempt to hold the insect firmly, he sticks a pin right through it and kills it. And thus "its voice vanishes with its life", and the man will never fully know the answer.

This fable has been extensively interpreted, including as an attack on Grassi's supposed "hard-handed" and inelegant method of reasoning. But it stirred Urban's heart because he felt it said something essential about all inquiry into nature: the deepest cause was in principle inaccessible to the human intellect. The ways of God were impenetrable. He had an infinite number of means at his disposal when it came to nature and its phenomena, and it was folly for man to claim that one particular explanation was absolutely true.

Urban VIII was therefore not especially fearful of Copernican theories. They might be of interest, plausible even – but they could never make claim to be the Truth, and therefore could not come into conflict with religion.

Galileo was of another persuasion. For him, the truth was one and indivisible, science and religion two sides of the same coin. But he well knew how Pope Urban thought – and he noted it for future use.

A Marvellous Combination of Circumstances

In his villa at Bellosguardo Galileo could review the situation and conclude that things had turned out advantageously. The uncertainty regarding the

Medici family that had crept in when Cosimo II had died, was resolved. Galileo had used his international network of contacts in the highest echelons of European society. He had meekly addressed himself to the Austrian Archduke Leopold, whom he had formerly presented with telescopes and observations about the tides. His application resulted in Leopold writing to his sister Maria Maddalena, mother of the still under-age heir apparent, Grand Duke Ferdinando II, warmly recommending that the court at Florence retain the services of Galileo. Ferdinando was in any case a mild and tractable young man, who showed no sign of distinguishing himself intellectually or in any other way.

However this was as nothing compared to "this marvellous combination of circumstances"[58] (*mirabil congiuntura* – Galileo's words in a letter to Prince Cesi) which had come about in Rome. The change of popes, the Lyncean Academicians' entry into the court of the Vatican and the propitious timing of the publication of *The Assayer* opened the way – not simply for a new personal triumph, but with a little care and luck, an evasion of the prohibition of 1616 and a new launch of the Copernican theory.

With some papal goodwill, Galileo might even resolve one of his personal problems. It was not impossible that Urban VIII, as a sign of his favour, might appoint young Vincenzio Galilei to a clerical sinecure that would give him a modest independent income. His son was making heavy weather of his studies at Pisa.

It was important to get to Rome again, to pay his humble respects to the Pope, but also to see how the intellectual land really lay now.

Galileo was suffering from constant ill-health, he had turned sixty in February 1624 and well knew that it was high time for him to bring together all his practical experiments and theoretical deliberations into one great, comprehensive work. Despite his growing contemporary fame, he had as yet not written anything that could compare with Copernicus' *De Revolutionibus Orbium Coelestium* or for that matter his competitor Kepler's *The New Astronomy* or his newly published *World Harmony*.

He set out in April. Prince Cesi, too, with his extensive web of contacts in Rome, knew how important this opportunity was, both for the promotion of science in Italy and the Lyncean Academy's reputation. Cesi owned a large estate at Aquasparta in Umbria. It was his favourite place, and there he spent long periods of time in scientific investigations of all kinds. He now invited Galileo to break his journey there for a fortnight, as the estate lay roughly half way between Florence and Rome. There the two of them could discuss the situation in Rome with all the city's complex groupings and alliances, and arrive at some plan of action for the offensive against Urban VIII Barberini. Cesi could hardly have forgotten Galileo's overenthusiastic performance during his previous stay in Rome.

Galileo arrived in Rome on 23 April, and the very next day he had an hour's private audience with the Pope. Urban VIII was friendly and sympathetic as before, he promised to look out for a position for Vincenzio and invited the mathematician to come again. Galileo had a total of six meetings with the Pope over the course of one and a half months in Rome.

He also nurtured other important connections, especially the Pope's influential nephew Cardinal Francesco Barberini and a German bishop, Zollern, who was worried because German Protestant scholars were now more than ever following in Kepler's footsteps and accepting the Copernican system. And so the Protestants were in the process of acquiring an ideological weapon against the papacy – and this in a situation in which the war between the Emperor's Catholic army and his Protestant subjects was being re-kindled on German soil.

However, with the passing of the weeks both Galileo and Cesi realised that despite the amicable reception, they had not got very far in moderating the terms on which cosmology could be discussed. Urban exuded friendship and respect – the Pope even wrote a warm letter of recommendation for Galileo to take back to the court at Florence. But the decrees from 1616 remained firm. It was indeed possible to write about the Copernican system as a hypothesis and a basis for calculation, but only as long as one explicitly distanced oneself from the idea that it represented a physical reality. Urban clung to his theologically based scepticism: God's ways could not be described fully by human intellect. And so ultimate proof that Copernicus was right could never be brought. It was not even theoretically possible to bring such a proof.

Galileo had his tides. He had no intention of relinquishing those. But for now he must return to Bellosguardo and think things over.

He did not think long. After consulting his friends in the Lyncean Academy he decided to fly a kite. The opportunity was there waiting for him – as it had been in fact for the past eight years.

During his previous visit to Rome in 1615–16, when Galileo had crushed his opponents in improvised discussions about Copernicus in the homes of the Roman ruling classes, he had bumped into an old acquaintance, Francesco Ingoli, who had studied law at Padua. Ingoli had chosen a career in the church, but he was interested in astronomy and had published a couple of minor works on heavenly phenomena. He was not convinced by Galileo's rhetoric and so he published a small paper, *A Disputation on the Location and Stability of the Earth*, in which he attempted to counter the Copernican doctrine.

Ingoli had tried to use physical and astronomical arguments against Galileo, not merely theological ones. Galileo was not especially impressed with these arguments, and it is not certain that he intended to respond to them. The events of 1616 rendered the question uninteresting – in the wake of the decisions by the Inquisition and the Congregation of the Index it would, to say the least, have been unwise to mount a public defence of Copernican ideas. But certain perverse opponents interpreted Galileo's silence differently: they thought that he had actually been refuted by Ingoli and had no defence to offer.

Now the situation had changed – or in any case Galileo and Prince Cesi *judged* that it had. With papal goodwill it should be possible to embark on a precarious balancing act: on the one side defending the theories of Copernicus against Ingoli 's arguments, on the other still declaring that the theory was *not* correct, because it, in turn, flew in the face of a theological understanding of the reality that existed on a different, superior plane.

The project could be launched in support of the Church: Protestants were not to be left under the misapprehension that Catholics were so stupid that they could not reason clearly and scientifically! On the other hand, their *piety* caused them to relinquish a theory that they had carefully provided proofs for, if it ran contrary to the Bible's express words and the Church's authority.

It was a complex task. Galileo did not spend long on the scientific parts of his "Letter to Ingoli", in which for the first time he explained in writing why the Earth could move without us noticing it in our everyday lives. The problem was the necessary adjustments in regard to theology. He sent a draft to Rome, where his friends made copies and suggested various corrections before the letter was published – or delivered to its addressee.

But this process was long drawn out and nothing was actively done with the letter throughout the entire winter of 1624–25, despite the fact that Ingoli, who had heard about the long delayed response, asked to see it. In fact, selected portions of the text were actually read out to Pope Urban. It was his trusted Secretary Ciampoli who had done this, and he could report back on His Holiness' uncommon goodwill. Just *which* extracts he had read he did not mention. They were likely to have been the less controversial ones.

Prince Cesi intervened in the late spring. He recommended that the "Letter to Ingoli" should not be given to its addressee, and certainly not printed. Things were afoot in Rome that he was not happy about.

War and Heresy

In September 1624, as Galileo sat at home in Florence finishing his "Letter to Ingoli", a new Jesuit professor was taking up his position at the Collegio Romano. His name was Father Spinola, and he used his inaugural lecture to launch a sharp attack on those who "sowed the seeds of heresy" by airing new and unbiblical scientific views. There is little doubt that he had discussed this with his religious brother, the offended Father Grassi.

Orazio Grassi had now had time to study *The Assayer* in considerable depth, and was ready to hit back at Galileo. Most of his colleagues were ready to support him. One of the points on which Grassi attacked, though excessively quibbling, did lie in an extremely dangerous area. It concerned Galileo's important distinction between a body's actual, fundamental characteristics, and its secondary ones, which Galileo almost regarded as a kind of illusion created by the senses.

So what of the Eucharist? Grassi asked quietly.

According to Catholic doctrine, the bread and the wine were physically *changed* by a miracle into Jesus' body and blood, although their outer characteristics – colour, smell and taste – remained the same. But Galileo said that colour, smell and taste were "merely indications". This must mean that he abjured the miracle itself – there could hardly be anything miraculous in preserving indications, illusions that had their root in the human sensory system.

Grassi was no small fry within the powerful Jesuit order. Just at that time he had been given the prestigious job of designing the Collegio Romano's new church, dedicated to the order's founder, the saintly Ignatius. The Sant' Ignazio church never turned out as grandly as planned, but that was not Grassi's fault. He intended to give it a magnificent dome – which perhaps not totally unintentionally would have blocked out the light to the library of the Dominican monastery close by!

Grassi's Eucharist objection – which was printed in his book *Comparison of the Weights of The Balance and The Assayer* a couple of years later – was regarded by most as a curiosity. Galileo's thoughts were purely scientific and never had any pretensions to theological relevance. Furthermore they were promulgated in a book that had passed the censor and was even dedicated to the Pope. But the objection *was* disquieting, because if it was sustained by the Inquisition, no supporter in the world would be able to save Galileo from a charge of heresy. And perhaps that was the hint behind Father Spinola's general railing against the sowers of apostasy.

Galileo was worried and made discreet enquiries. Privately he was assured that no steps would be taken against *The Assayer*.

But his fears were not groundless. Simultaneously he had news from Rome that must needs worry him: the hunt for heretics was continuing under Urban VIII as well.

Just before Christmas 1624 a disturbing event took place not far from the Collegio Romano. In the Dominicans' church of Santa Maria sopra Minerva the members of the Inquisition had assembled to pass a sentence. The accused was the former Archbishop, Marco de Dominis. He had once dabbled in geometry and optics, lived in the Republic of Venice and had been a friend of Paolo Sarpi and Gianfrancesco Sagredo. As an enemy of Bellarmine and the Pope's increasing worldly power he had fled to England, where he eventually oversaw the publication of Sarpi's great work on the Council of Trent, a work that depicted in such detail the intrigues behind the many far-reaching decisions that it was immediately placed on the *Index*.

But de Dominis quarrelled with the English and returned to Rome where he abjured all his heretical acts (including the book publication). With his dissolute background he was a social success in Rome during the first, optimistic year of Urban VIII's pontificate. He resumed his scientific work and wrote a dissertation on Galileo's pet subject, the tides, which he had had every opportunity of studying on the English Channel.

But Marco de Dominis' enemies had most decidedly not forgiven him. He was arrested, his effects examined, and a charge was then brought against him concerning heretical things he had written about marriage in an unpublished manuscript. During his examination he admitted that he drew a distinction between two classes of religious dogma. Those which concerned faith directly were sacrosanct. But others, for instance a number of resolutions adopted at Trent, could be discussed. This was precisely the same distinction that Galileo had drawn for interpreting cosmological phenomena.

Archbishop de Dominis was sentenced to death. It was a sentence that aroused a great deal of attention in Rome, as the accused was already dead when the sentence was read. He had passed away in prison during the trial; poisoned, according to some. Nevertheless, the Inquisition spared no pains when once the judges had finished their investigations. The body was dug up and lay in a coffin in the "dock". After the death sentence had been passed, de Dominis' cadaver and all his writings were driven from Santa Maria sopra Minerva to Campo de' Fiori, where they were all publicly burnt, together with a portrait.

Father Spinola and Father Grassi were not Galileo's only opponents amongst the Jesuits. Towards the end of the year a German Jesuit arrived in Rome. He was an astronomer, a thorough observer, who had studied sunspots minutely for many years. He well knew that the constrictions which a worship of Aristotle and Ptolemy placed on the investigation of natural things had to go. Some said that deep down he had become a Copernican, a fact which he, of course, could hardly announce from the lectern in the Collegio Romano.

This Jesuit was Father Christopher Scheiner, the sunspots observer who had tried to build up a good relationship with Galileo, but had only succeeded in being blackguarded twice by Galileo's pen. And if Spinola was fearful and Grassi insulted, Scheiner was furious.

Scheiner quickly got his bearings in the intellectual and clerical landscape of his Roman religious brothers. He gained much influence over a cardinal who had just joined the Jesuits. This was Alessandro Orsini, the man who had attempted to teach Paul V about the tides on the same day that the Inquisition condemned Copernicus. Now, even his sympathy had swung away from Galileo.

Just then, another of Galileo's friends, the Pope's nephew, Cardinal Francesco Barberini, disappeared for a while from Rome and the papal sphere of influence, because he had been appointed diplomatic envoy to Paris. Conditions in Rome were always shifting in this way, alliances were formed and dissolved, sympathies radically re-thought.

There were sufficient grounds for Prince Cesi to call for some caution, despite the marvellous contacts the Lynceans still had with the Vatican. And even now the greatest revolution of all had not made itself much felt, because it came gradually, at first imperceptibly. Of all the changes it was certainly the most important, because it occurred right at the top of the system: Pope Urban VIII himself was in the process of transformation.

European Power Struggle and Roman Nephews

The Thirty Years War was the first great European war. It began as a struggle over power and religion in Bohemia, and was fought out mainly in the innumerable German states, large and small – there were approximately

300 independent political entities of which eighty were large enough to play a practical part. But gradually the balance of power in Europe itself became the real driving force behind the bloodbath. Spanish influence at the Imperial court at Prague was strong, and the Spaniards could legitimise their struggle for power as support for the Emperor and as a campaign for Catholicism in apostate northern Europe.

The energetic Ferdinand II became emperor in 1619, and he immediately put all his resources into gaining control of the Bohemian Protestant rebels. At first the war went extremely well, from Rome's point of view too. The Protestant forces were crushed in 1620 at Bila Hora – the white mountain – in Bohemia. With Spanish help Emperor Ferdinand brutally followed this up by confiscating estates and executing prominent Protestants in Prague.

After that he organised a large army under the command of the tactical genius Wallenstein. Wallenstein solved the logistical problems of war by simply allowing his soldiers to plunder and pillage their way through the landscape – doing away with the need for long supply lines. This method also had the advantage that it eased army recruitment. Within the plundered areas there was simply no other way to live.

The success of the Emperor and Wallenstein in the south meant that they began turning their gaze to northern Germany. Not only might they win back the Lutheran areas to Catholicism, but a naval base on the Baltic would mean influence over the North Sea and the Baltic States.

This frightened German Hanseatic cities as well as the English, Dutch, Swedes and Danes. But it was also very worrying for Catholic France, which certainly did not want its arch-rival Spain to gain a dominant position in large parts of Europe.

Thus the war put the Pope in a dilemma. The papal court was traditionally the arena for intense rivalry between Spain and France. But the Pope was also a temporal monarch who ruled over the Papal States, and he had the Spanish dominated Kingdom of Naples as a powerful neighbour right on his southern border.

Urban VIII Barberini was a francophile. His career had taken off in France, where he had shone at court. As pope he needed a powerful France for political reasons, as a counterweight to Spain. But the French, led by Cardinal Richelieu gradually began to support the Protestants quite openly. This was something that the Prince of the Church could not countenance. Urban *had* to show solidarity with the imperial war effort, but his solidarity was limited to words of encouragement: he provided neither funds nor troops.

Instead the money went to the construction of the Papal States' own defences, and not a little to the Barberini family itself. The extent of this nepotism can be glimpsed from the fact that when he was elected pope, Maffeo Barberini had an estimated fortune of 15,000 scudi. After a few years on St Peter's throne he purchased an entire region and its noble title for his nephew Taddeo for 750,000 scudi.[59] And Taddeo was only one of many relatives.

This caused discont in Rome. People whispered that the Pope was not sufficiently concerned with the Catholic faith and its dissemination – indeed, that he was apathetic or veering towards the heretical.

Urban VIII noted the change of atmosphere and became more and more mistrustful. His open, inquisitive nature slowly congealed into a rigid self-importance that brooked no contradiction or criticism, either of his political or religious judgements or his purely personal and not inconsiderable vanity.

The Barberini Pope had always been superstitious, something Galileo's enemies in Rome tried to capitalise on. There were rumours of a horoscope that predicted imminent death for both Urban and his nephew Taddeo. The horoscope was said to have been cast by a Vallumbrosan monk, and some (perhaps knowing that Galileo had gone to school at Vallombrosa) claimed that it was actually the "mathematician and astrologer" Galileo who was responsible. Galileo understood the gravity of this and got one of his friends in Rome to intercede, a Florentine with the resounding name of Michelangelo Buonarroti, nephew of the great Renaissance master. Galileo extricated himself from the affair – but the danger of rousing the Pope's displeasure was emphasised by the fate of the Vallumbrosian monk: he was arrested and died in prison awaiting his trial.

Urban VIII had no need to fear the portents of the sky; he was to live another fourteen years. But increasingly often he was away from Rome. He had a magnificent papal summer residence built in the Alban Hills some miles south of the city, where the summer climate was pleasant and the white wine excellent. Out here, near the small town of Castelgandolfo Urban felt relatively safe from his adversaries, but he still had his food tasted by servants before he dared to eat it himself.

Within Italy itself problems were mounting. The Pope's relations with Tuscany and the grand ducal family took a serious turn for the worse due to an inheritance conflict over the little dukedom of Urbino. For safety's sake Urban had his troops occupy the area, which he wanted to annexe to the Papal States.

All this occurred gradually during the 1620s. The emergence of a new epoch, the "marvellous combination of circumstances" which Galileo had rejoiced over in 1623, was definitely in the process of receding.

The Old and the New

At home at Bellosguardo Galileo followed events in Rome. But he did this second hand, via his correspondents and did not entirely register the change of atmosphere that was slowly taking place. In addition he was frequently ill, or concerned with family matters.

His brother Michelangelo, who was still a musician in Germany, sent his entire family – wife, eight children and nursery maid – away from poverty and war, and home to Galileo and the security of Tuscany. They lived at Bellosguardo for a good year and filled the house with more life than Galileo really appreciated.

Then there were his daughters in the convent of San Matteo. They needed constant visits and help when they wanted contact with the outside world. A ceaseless stream of small gifts passed between father and daughters – and when Galileo was absent from Florence, they kept in contact by letter.

His boy Vincenzio was still a worry. When Pope Urban finally found a sinecure for Galileo's son, he immediately turned it down because he did not want to accept any support from ecclesiastical quarters.

Taken as a whole the reports he was getting from Rome were still quite optimistic in tone. Above all the Lynceans had maintained their influence at the papal court. Prince Cesi was highly respected, and Ciampoli was still the Pope's Secretary. He could confirm that Galileo personally enjoyed His Holiness' high regard. Father Grassi's attack on *The Assayer*, with its ominous hint that Galileo's view of natural philosophy was incompatible with the Catholic understanding of the Eucharist, was finally printed in 1626. The attack was probably of little consequence (but see postscript p. 204). Grassi's book was published in Paris, probably on grounds of discretion because *The Assayer* was, after all, dedicated to the Pope. Anyway, Galileo did not much concern himself with Grassi's answer – he regarded the comet debate as at an end.

Despite his problems, Galileo now decided to risk setting out his theory, and hazarding everything on the tides. He wanted to write the great, definitive work that no one would be able to surpass; a unified account in classical dialogue form, in Italian, just as his father half a century before had written *his* great work *Dialogue on Ancient and Modern Music*. It was precisely the "ancient and modern" that Galileo also wanted to discuss: the Ptolemaic system versus the Copernican. Within the dialogue the certain arguments in favour of Copernicus would be advanced, accompanied naturally by the reservations necessitated by the Church's attitude. Galileo believed he had Pope Urban's permission for such a "contingent" discussion.

But he was working alone at Bellosguardo. His contact with others was limited largely to discussions and correspondence with people who agreed with him. Galileo had no great patience with those who refused see that Copernicus was right, and his patience did not increase as he wrote his way through the Copernican arguments. As the work progressed, his reservations and provisos became fewer and more brief – the reservations that were to show that he "only" regarded the idea of the Earth's motion as hypothetical speculation.

Galileo had long known what he would call his great work. Its title was to be *Discourse on the Ebb and Flow of the Sea (Dialogo del flusso e deflusso della marea)*.

The writing progressed slowly and with long pauses. His family demanded their share of his attention the whole time. However, they were not only a source of problems: Vincenzio had at last finished his studies at Pisa, and had come home to Florence with a degree in law. Immediately after this he got engaged and was married in January 1629. Galileo's first grandchild was born in December of that same year and was named after him.

The university of Pisa wanted to get Galileo off its payroll, as his employment there was a pure formality and meant little more than that the university had to pay his salary. Galileo mobilised the young Grand Duke, who eventually saw to it that the contract that his father Cosimo II had drawn up, was respected. But such things took time and energy.

On Christmas Eve 1629 – after an intensive period of work – he wrote to Prince Cesi mentioning a new, serious health problem: Galileo the observer, the telescope virtuoso, the mathematician with the eyes of the lynx, was slowly losing his sight.

But his greatest difficulties lay in the actual realisation of his project. He came across a good many problems which he had to reconsider, especially in connection with tides, which without doubt *were* difficult to solve. The dialogues were also to elucidate various phenomena connected with motion, so it was necessary to go through all the old material he had, right back to his time at Padua. Furthermore, these observations often produced interesting digressions which were highly suited to the dialogue form and imparted life to his account. But it took time to put it all into words.

While the work on the tides and their significance slowly took shape under Galileo's pen in Florence, work was also going forward in Rome. The German Jesuit, Father Scheiner, wished to publish his meticulous observations of sunspots, but he was also very keen to get in some good swipes at Galileo.

Scheiner's feelings about Galileo are reminiscent of spurned love, which turns to a kind of hate. He had *tried* to establish an intellectual dialogue with his *Mathematical Discourses*, but had caught Galileo on what was perhaps his most sensitive spot, the prestige associated with priority of discovery. After that, Scheiner was the recipient of Galileo's contempt – or at least that was the way he viewed it himself.

Scheiner's work was a massive volume of 784 double column pages. The "First Book" filled the introductory 66 pages, and was largely a sustained attack on Galileo. Scheiner asserted his right as the original discoverer of sunspots, prior to Galileo and quite independently of him.

And the Jesuit astronomer did not stop there. He maintained that Galileo had not even noticed that the spots described curved trajectories above the Sun's surface, and that he had not actually discovered that the spots were surface phenomena and that the Sun turned on its own axis. If Galileo *had* ever written anything like this, it was nothing but sheer luck and guesswork!

The movement of the spots was a very important matter, and Scheiner knew it. As a Jesuit he could not openly discuss just *how* important it was. But he used the remainder of his work to describe this and other solar phenomena in precise detail. Furthermore he sharply criticised the traditional Aristotelian background to astronomy, especially the doctrine concerning the heavens' immutability. This was an attempt to liberate Jesuit science from the straightjacket of Aristotle – but he was unable to follow this up by throwing off Ptolemy's at the same time.

Scheiner's book was called *Rosa Ursina*. The title was a tribute to his patrons, the Orsini brothers. Prince Paolo Orsini, brother of Cardinal Allesandro, had even taken care of the printing, without presumably having read the manuscript with its hefty attacks on Galileo. In any event, the Grand Duke's mathematician received an apology when he wrote a letter of complaint to the Prince.

Galileo probably read the entire book, despite the attacks on him at the beginning, as he had good use for its precise descriptions of sunspots' movements in his own book, which was now nearing completion.

On 1 May 1630 Galileo went to Rome for the fifth time in his life. As usual he stayed at the Ambassador's residence, the Villa Medici. He had with him the manuscript of *Discourse on the Ebb and Flow of the Sea*. The book was to be printed by the Lyncean Academy, but it was necessary for the author himself to take part in the process leading to the Church's approval – for everyone realised that Galileo's ebb and flow swept across dangerous and muddy waters.

Not only was it quite apparent that he had many powerful enemies in Rome, but he also had to keep to the strict prohibition of 1616: "that the Sun is the centre of the world" was foolish and heretical, "that the Earth moves according to the whole of itself, also with a diurnal motion" was poor philosophy and incorrect creed. It could not be denied that Galileo's voluminous manuscript dealt for the most part with exactly these two subjects.

The solution was to present them as hypotheses, devices, calculational examples.

Galileo had only one audience with Urban VIII this time, but their meeting was pleasant enough, even though the Pope repeated his favourite thesis that all theories were in principle unprovable in the light of God's omnipotence.

The Pope did however distance himself discreetly from the decree of 1616. At least, that was the impression given to a man who had discussed the problem with him in March, a character who now, improbably enough, found himself in Rome and close to the papal court: Tommaso Campanella. That rebellious Dominican had been freed from gaol in Naples in 1626, on the initiative of Urban himself. He was first handed over to the Holy Office in Rome, but in 1629 he gained his complete freedom.

As previously, Campanella wanted open discussion on all cosmological points of view, and referred in a letter to a private comment by Urban VIII to the effect that, if it had been up to him in 1616, there would have been no injunction. In the meantime Galileo – to the enthusiastic monk's bitter disappointment – continued to keep his distance from his ardent admirer, even though Campanella had been partially rehabilitated.

The papal censor in Rome bore the impressive honorary title of "Master of the Sacred Palace". His name was Father Riccardi, and he now had the responsibility of reading and then potentially approving Galileo's manuscript. Father Riccardi came from Florence and was a relative of the Tuscan Ambassador's wife. He had earlier read *The Assayer* with great pleasure. Riccardi also knew, of course, that the Pope looked kindly on the Grand Duke's mathematician, and had done so for many years.

But even though Riccardi was positive to begin with, his doubts grew as he read. The *Discourse on the Ebb and Flow of the Sea* promulgated the teachings of Copernicus with great fervour, persuasive force – and stinging irony directed at those who remained stuck fast in Aristotle and Ptolemy. Certainly, it ended by stating that nothing was certain and that Copernicus' doctrine should only be regarded as a hypothesis, but this conclusion seemed to be an unconvincing afterthought to say the least.

Father Riccardi could not take responsibility for it in its present form. He insisted on a new introduction, a clearer summary and the correction of various minor points. The main thing was that the condemnation of Copernicus' book by the Congregation of the Index should not be made to appear ludicrous, but rather as a sensible decision.

Riccardi asked one of the monks, who was a mathematician, to look through the manuscript and make corrections. But the mathematician did not find much to correct. He realised, privately, that Galileo was right, and looked forward to a new discussion on what was acceptable cosmology.

This was not a lot of help to Father Riccardi. He was under pressure from Galileo's influential friends, and grudgingly agreed to give the book provisional approval, on condition that Galileo himself went through the manuscript again and sent the corrected pages to him as they were finished. This enabled the laborious work of typesetting and printing to begin.

For safety's sake – and this shows just how important the matter was – he took the question up with Pope Urban directly. The Pope was satisfied with Riccardi's explanation and gave the go-ahead, but he had one reservation: the title.

Discourse on the Ebb and Flow of the Sea sounded very innocent. But one thing Urban VIII was quite certainly aware of – not least because of their many conversations – was that Galileo did not view the tide as an *argument* that could be marshalled in favour of a Copernican hypothesis, but as an irrefutable, physical *proof*.

Papal intervention saved Galileo from the historical ignominy of his magnum opus bearing for all posterity a title that testified to his gravest error. What he thought about the matter is unknown. Urban suggested instead *Dialogo sopra i due massimi sistemi del mondo* or something in that vein.

Dialogue Concerning the Two Chief World Systems. On 26 June 1630 Galileo travelled home from Rome with a new title for his work, convinced that everything was now in order. He only had to go through it once again, clear up any minor problems with the censor, and maybe add something of importance on the movement of sunspots. After that he would send it back to Rome, where Prince Cesi would take care of the printing on behalf of the *Accademia dei Lincei.*

"An Advantageous Decree"

While Galileo was just starting his revision, something catastrophic occurred in Rome. The unifying force behind his varied and diverse connections in the city, was suddenly gone. The founder of the Academy, the undisputed leader of the sharp-eyed lynxes, Prince Frederico Cesi, died quite suddenly on 1 August, only 45 years old.

Cesi left no will behind him and no adult heirs. As he was the organisational and financial force behind the Lyncean Academy, all its work was paralysed. No one else could make the necessary decisions, and the remaining members were forced to concentrate on one immediate practical problem: they had to save the Academy's library. There were books in it that would not bear close scrutiny by the Church authorities.

Galileo was left with his shock and sadness at the loss of a close friend, a wise enthusiast and an energetic worker. In addition he had a finished manuscript on his hands – but now no publisher.

Nor was the *Dialogue* a straightforward manuscript that could simply be turned over to a book-printer. He realised that the best thing would be to get the book printed in Florence, where he personally could oversee the process. But this brought with it a new problem: Father Riccardi's provisional approval was only valid in Rome.

The "Master of the Sacred Palace" was able to issue a general permission to print anywhere, but for this he stipulated a condition: he wanted the manuscript so that he could go through it himself once more, and for the sake of caution with the Lyncean Ciampoli, Galileo's friend and the Pope's Secretary.

Father Riccardi was clearly in a difficult bind. He was a Dominican, and he knew of course that powerful forces amongst the Jesuits in Rome were out to get Galileo – and that they might not worry over much about bringing down a Dominican at the same time, considering the traditional tensions between the two orders. On the other hand, Galileo *was* a favourite of the Pope, and he could also rely on the Grand Duke of Tuscany, still a powerful force on the Italian mainland.

But early that autumn the plague struck northern Italy, right down as far as Tuscany. The Papal States introduced strict quarantine rules, which even applied to large packages. This made it difficult to edit the manuscript in Rome. Galileo asked to be allowed to make the revision in Florence, and just send the introduction and conclusion to Rome for final approval. Riccardi, sensing problems, tried to spin the matter out.

A whole year passed like this to the ageing Galileo's great dismay. He mobilised the Grand Duke's Secretary and his Roman Ambassador, and in the summer of 1631 he received Riccardi's very grudging licence to print, his *imprimatur*. Although in fact no such approval really existed. What happened was that Riccardi sent instructions to the local Inquisitor in Florence, together with a draft preface that *had* go into the book, if not in those exact words at least with the same content.

This complex and lengthy process helped, if nothing else, to obscure the responsibility for the approval of the book, not for its content. *That* rested in the final analysis with the author.

With Riccardi's provisional approval the printing could at last begin, but even that took an inordinately long time. The *Dialogue* did not come out until 21 February 1632, with a dedication to Grand Duke Ferdinando II. The dedication is only two, short, sober pages long – quite bereft of the bombastic, high-flown language that characterised the introduction to *The Starry Message* twenty years earlier. Galileo makes the point that, in philosophy, one man's insight is worth more than a thousand men's opinions – provided that insight is correct. He says nothing more, but lets the reader decide how this maxim should apply to Ptolemy and Copernicus. His tribute to the Grand Duke is limited to what was surely a deep-felt gratitude for his financial help, together with an apposite comment about how it was through Ferdinando's agency that the book was finally printed at all.

The preface follows. The introduction is here given in its entirety:

> "Several years ago there was published in Rome a salutary edict which, in order to obviate the dangerous tendencies of our present age, imposed a seasonable silence upon the Pythagorean opinion that the earth moves. There were those who impudently asserted that this decree had its origin not in judicious enquiry, but in passion none too well informed. Complaints were to be heard that advisors who were totally unskilled at astronomical observations ought not to clip the wings of reflective intellects by means of rash prohibitions.
>
> "Upon hearing such carping insolence, my zeal could not be contained. Being thoroughly informed about that prudent determination, I decided to appear openly in the theatre of the world as a witness of the sober truth.

> I was at that time in Rome; I was not only received by the most eminent prelates of that Court, but had their applause; indeed, this decree was not published without some previous notice of it having been given to me. Therefore I propose in the present work to show to foreign nations that as much is understood of this matter in Italy, and particularly in Rome, as transalpine diligence can ever have imagined."[60]

Galileo certainly had been given "previous notice of the decree" in 1616! This was at his meeting with Bellarmine which ended in a strict warning not to portray the teachings of Copernicus as physical truths, and with an even clearer – probably downright threatening – reminder from Cardinal Segizzi. But Bellarmine and Segizzi were both long since dead.

Galileo's preface could be seen as a masterpiece of intelligent self-restraint. He spoke from an impregnable position as defender of Catholic intellectuality, while at the same time bending to the religious commands of a higher order. The preface – especially in the light of the rest of the book – could however also be seen as ironical hypocrisy. It all depended on the reader.

Two Wise Men – and a Third

Three men meet in a Venetian palace. They have come together to discuss "God's wonders in the heavens and on Earth", more specifically which of the two competing "world systems" is right: the Ptolemaic or the Copernican. As truly inquisitive men they have set aside four whole days for their discussions.

The owner of the palace is called Sagredo. He is wise and well informed, not a specialist in science or philosophy, but quick-thinking and well acquainted with the range of positions and views. He has invited a representative of each of the two world philosophies: the Copernican, Salviati, and the solid Aristotelian, Simplicio.

This is the literary structure, the fictional framework if you like, round Galileo's main work. Within it he tried to create a context in which all thinking Italian readers could *themselves* take a position about the degree of truth in the cosmological discussions, without being distracted by any ecclesiastically conditioned interpretation.

In order for this to succeed, it was not only the scholastic material that had to be convincing. The literary structure had also to grip the reader and preferably hold him captive throughout the lengthy book.

The author in Galileo brings this off. Even though the conversation between the three of them sometimes necessarily assumes the character of a string of deductions, it always remains a *conversation*. The three are individually drawn, each has his "voice", and they are certainly not reluctant to contribute quick, witty comments and characterisations. After Simplicio has carefully described how all the material in the heavens is unalterably and impenetrably solid (because Aristotle says so), Sagredo exclaims: "What excellent stuff, the sky, for anyone who could get hold of it for building a palace!" But Salviati disagrees:

> "Rather, what terrible stuff, being completely invisible because of its extreme transparency. One could not move about the rooms without grave danger of running into the doorposts and breaking one's head."[61]

Like all writers Galileo took elements of his characters from himself.

The author Italo Calvino has pointed out that Salviati and Sagredo represent two aspects of Galileo's personality: Salviati stands for his careful, methodical reasoning, while Sagredo uses his imagination, draws unexpected conclusions, asks surprising questions: what does life on the moon look like, if it exists? What would happen if the Earth *stopped* dead in its Copernican revolutions?[62]

Simplicio, by contrast, is no worthy opponent. He is constantly portrayed in a comic light with his credulous references to Aristotle and his commentators, not to mention contemporary anti-Copernicans. He thinks sluggishly and needs to have Salviati's reasoning thoroughly explained, whereas Sagredo grasps it immediately and often adds perceptive comments. When the others ask if he has read *The Assayer* or *Letters on Sunspots*, Simplicio answers that he has flipped through them, but has spent most of his time on more solid studies.

There is a fourth person mentioned in the book, but never by name, he is simply called "our mutual friend" or "the academician". This is Galileo himself, and Salviati, it must be said, refers to *him* almost as Simplicio does to Aristotle.

Clearly, the names are not accidental. Sagredo was Galileo's Venetian friend and benefactor from his years at Padua, Salviati the rich Florentine who owned the Villa delle Selve where Galileo had often lived and worked. He says in his preface that he wants the reputation of these two deceased friends to live on in the pages of his book. Simplicio, on the other hand, is a kind of pseudonym – it stands as it were for the average Aristotelian philosopher: "... whose greatest obstacle in apprehending the truth seemed to be the reputation he had acquired by his interpretations of Aristotle," as Galileo says in his preface.

The actual *name* Simplicio was in fact taken from a well-known sixth century Aristotelian commentator. But it certainly was not adopted at random – the Italian word *semplicione* means "unsophisticated person".

The first day's discussions concentrate largely on the relationship between earthly mutability and heavenly perfection. Poor Simplicio is bombarded with information about comets, sunspots and the Moon. He even has Aristotle turned against him, when Salviati ironically notes that it must be far better Aristotelian philosophy to say "Heaven is changeable because my senses tell me so" than "Heaven is immutable because Aristotle worked it out".

As the book unfolds Simplicio – and the reader – are treated to various lessons in the theory of motion, astronomy and optics. The unfortunate philosopher must unwillingly admit that there are one or two things he has not understood – but he defends himself stoutly with the aid of an impressive array of authorities old and new. One such is "a recent little book of hypotheses", which is supposed to refute all Copernican claims.

This book is Father Scheiner's *Mathematical Discourses*, the little book that the Jesuit had once sent to Galileo many years earlier in the hope of provoking a reply and a discussion. Now he receives his answer – it is much delayed but, to make up for it, pretty clear. To take just one example of Salviati and Sagredo's comments: they assume that the author (who is not mentioned by name) cannot be so foolish as to believe what he himself has written, but is trying to hoodwink people. And as if that were not enough:

> "Those who have nets to snare the common people know also how to be the authors of other men's inventions, so long as these are not ancient ones and have not been published in the schools and in the market places so they are more than familiar to everyone."[63]

In other words it is still the priority to the discovery of sunspots that rankles here.

Towards the end of the first day, Salviati makes some comments on the relationship between human and divine understanding. He says that it is true that human knowledge is nothing compared to God's, for the latter's is infinite, and even something is nothing compared to the infinite. But as regards the few things about which man can achieve true knowledge, his knowledge is *qualitatively* as certain as God's, if it is underpinned with definite proof – there is no extra degree of certainty above and beyond that which can be demonstrated incontrovertibly. This only applies to limited aspects of arithmetic and geometry, but the assertion still causes Simplicio to exclaim:

> "This speech strikes me as very bold and daring."[64]

Far from it, Salviati replies – these are perfectly ordinary statements. But on this one point unfortunately, Simplicio is right.

The discussions on the second day take up the largest part of the book. Here, the most difficult part of the Copernican theory is aired. If the Earth really does revolve completely on its own axis in the course of a day, how can it be that we who live there, do not experience the least sensation of it?

Galileo was quite used to meeting such arguments in discussions, and his elucidation is therefore a pedagogic masterpiece. Taking dozens of examples from daily life, Salviati hammers home the tenets of motion. The most important of these is that all motion is *relative*. When we are on a ship travelling at a constant speed, we only notice motion in relation to the water, to islands, other boats etc. – *not* in relation to other objects on the ship, which are moving at precisely the same speed as ourselves. It is the same with the Earth, because the planet and everything on it, including us human beings, are making the same journey.

Salviati's Copernican defence is so full of power and conviction that it feels almost painful when Galileo suddenly recalls the conditions under which he is writing. Then, quickly, he interjects a little aside

> "[I] who am impartial between these two opinions and masquerade as Copernicus only as an actor in these plays of ours..."[65]

But it gets worse. On the third day the discussion becomes more technical and astronomical, dealing mainly with the Earth's annual movement around the Sun, Copernicus' quintessential point: which heavenly bodies move and which stand still. Painstakingly, Salviati describes all the seemingly extraordinary phenomena Ptolemy must explain, but which vanish if one reverses the system and gives the Earth an orbit:

> "The illnesses are in Ptolemy, and the cure for them in Copernicus."[66]

This is all to do with the planets and the so-called retrograde motions.

But here, too, sunspots are dealt with. And as Salviati makes clear:

> "our Lyncean Academician" discovered them in 1610, in Padua. Furthermore, he "spoke about them to many people here in Venice, some of whom are yet living."[67]

This, to put it mildly, is definitely a lapse of memory.

In a letter from Galileo to Maffeo Barberini dated 2 July 1612 he wrote that he saw the sunspots "about eighteen months ago", so about New Year 1611; not when he was living in Padua in the spring of 1610. The difference of nine months in the new dating may seem insignificant – but it is just enough to pre-empt Scheiner in the discovery. This is then thoroughly rubbed in. Galileo is, according to Salviati:

> "The original discoverer and observer of the solar spots (as indeed of all other novelties in the skies)."[68]

Sunspots describe what looks like gently curving trajectories across the face of the Sun. If one assumes that the Earth moves in a plane that is not absolutely vertical to the Sun's axis, the sunspots' motions will appear just like this from the Earth. This is presented as an argument in favour of Copernicus, based on observations Galileo was said to have made. There is no trace of these observations in Galileo's notes. Of course, it is possible that they have been lost, but what is certain is that *Scheiner* published just such observations in *Rosa Ursina*. It is not surprising that, having read the *Dialogue*, he believed that Galileo had simply used his own painstaking work of many years.

Then Salviati begins to demolish Scheiner's various anti-Copernican arguments. This is done without mention of his name or that of his book, but with phrases like "apish puerelities",[69] "trifling tomfooleries",[70] "gigantic fallacy".[71] It all culminates in Salviati addressing Scheiner directly: "O foolish man!"[72]

On the fourth day the three debaters finally tackle the tides, the original theme of the *Dialogue*. This chapter is shorter than the others and lacks their supple digressions; it chiefly consists of a continuous lecture from Salviati in which he lays out Galileo's complicated reasoning concerning the interplay between the Earth's rotation and its orbit in space. Kepler is given a friendly kick for his medieval superstitiousness about the Moon's influence – but otherwise the tone is much more subdued. It is as if part of Galileo *senses* that his much beloved tidal theory really is not convincing enough to crush its opponents.

But then the discussion must be brought to a close. And it cannot happen in the way that all of the preceding pages – close on 500 of them – have been building up to, with the obvious conclusion that everything supports Copernicus: the Sun is motionless, the Earth moves in its orbit and on its own axis. Quite the opposite, as Salviati suddenly says:

> "I do not claim and have not claimed from others that assent which I myself do not give to this invention, which may very easily turn out to be a most foolish hallucination and a majestic paradox."[73]

And so Simplicio is left with *his* conclusion:

> "Keeping always before my mind's eye a most solid doctrine that I once heard from a most eminent and learned person, and before which one must fall silent, I know that if I asked whether God in His infinite power and wisdom could have conferred upon the watery element its observed reciprocating motion using some other means than moving its containing vessels, both of you would reply that He could have, and that He would have known how to do this in many ways that are unthinkable to our minds."[74]

Both the others are in heartfelt agreement. And how could they be otherwise? For the eminent and learned person whose pet argument is here being rehearsed by the play's Pantaloon, is His Holiness himself, Galileo's intimate friend of many years, Urban VIII Barberini.

The Inquisition's Chambers

On 8 March 1632 a violent and scandalous episode took place in the Vatican. Before the entire College of Cardinals the leader of the pro-Spanish faction, Cardinal Gaspare Borgia, read out a sharp protest against the Pope and his lack of support for the Spanish war against the Protestants in Germany. Borgia took the unheard of step of hinting that a meeting should be convened to consider whether the Pope really had the necessary will to defend the Catholic faith.

Pope Urban and his faithful nephew Francesco tried to hush the rebellious Cardinal, without success. Finally, Urban's brother Antonio (who had also been appointed a cardinal) rose to restrain Borgia by force, but another worthy cleric held him back. The chamber was in uproar. One cardinal broke his spectacles, while another got so irate that he tore his biretta to shreds. Urban VIII had to summon the Swiss Guard to restore order in the assembly.

The cardinals left the room at the sight of the hefty guards with their halberds. The Pope was left, scarred, incensed – and politically weakened. He wanted to send Cardinal Borgia away from Rome, but dared not fearing that Spain, through the Kingdom of Naples, would intervene militarily. In a fit of what seemed like paranoia, he also imagined that the Grand Duke of Tuscany was making ready his fleet to put out from Livorno and attack the Papal States' harbours at Ostia and Civitavecchia. The purported excuse for this was the dispute between Tuscany and the Pope over the right to the Dukedom of Urbino. If the Pope's mind had been less distracted, he would have realised that the amicable Ferdinando II entertained no such warlike plans at all; on the contrary, he was more concerned for his people's welfare during the ravages of the plague epidemic.

Urban VIII saw that he had to display a stricter and more orthodox attitude if he were to retain his authority and protect himself from out-and-out scandals like Cardinal Borgia's outburst. But he did manage to exact a small revenge. He banished two less important cardinals who were also known to be pro-Spanish. These two had something else in common – they were close friends of Urban's own Secretary, the old Lyncean Giovanni Ciampoli.

And just to show that he was in earnest he got rid of Ciampoli too.

Ciampoli was – like all the members of the *Accademia dei Lincei* – a man with great gifts. But he was only too well aware of them, and often appeared rather arrogant to the people about him. When the Pope suddenly sent him away after many years' service, the explanation given in Rome was as follows: Urban, who was a poet, and very proud of the fact, wanted to write a personal pastoral letter in Latin. He showed a draft of this to several trusted colleagues, including Ciampoli. But his Secretary did not return it with the customary dose of apposite praise. Instead, he pulled the Pope's words apart and wrote a new, thoroughly reworked version.

It is possible that such an episode could have been the causal factor. But Ciampoli's links to pro-Spanish factions was probably more important. In any event, the Secretary got the sack from an infuriated Urban VIII.

During this spring of crises, the first copies of *Dialogue Concerning the Two Chief World Systems* arrived in Rome. The Pope had no time to immerse himself in it immediately, but there were sufficient eager readers – so many, indeed, that not all of them could get hold of the book.

One of the first was Galileo's indefatigable admirer, the pardoned prisoner and Dominican, Tommaso Campanella. He was very impressed with both the book's style and content, but he was not satisfied with the explanation concerning the tides and he plainly said so in letters to Galileo. He also added dryly: "Apelles will complain a lot about this book."[75]

"Apelles" was Father Christopher Scheiner's old pseudonym from their first discussion about sunspots. And Campanella was absolutely right. An eye-witness told of an episode in a bookseller's where Father Scheiner heard another priest praise the *Dialogue* as the best book ever published:

> "[Father Scheiner] was completely shaken up, his face changing colour, and with a huge trembling of his waist and his hands, so much so that the book dealer, who recounted the story to me, marvelled at it; and furthermore told me that the said Father Scheiner had stated that he would have paid ten gold scudi for one of those books so as to be able to respond right away."[76]

During the course of May and June more copies arrived in Rome. Scheiner got one. So did all of Galileo's other enemies in Rome. And one of them made certain that His Holiness was thoroughly informed about the rebellious and *unorthodox* book Galileo had written – the man all Rome knew enjoyed a very special position with the Pope.

Urban VIII had to show his authority, and he did not wait until he had read the book. The first to find himself in difficulties was the unfortunate Father Riccardi, "Master of the Sacred Palace", who after many ifs and buts had given the all clear for the printing in Florence. Riccardi was given to understand in no uncertain terms that he had failed in his duty: serious objections could be raised against the *Dialogue* in its present form.

Luckily for him, Riccardi could push much of the blame on to the Inquisitor in Florence. At the end of July he wrote what was, in the circumstances, a calm and courteous letter to Florence, explaining that Galileo's book had run up against problems in Rome and that it would be necessary to make alterations to it. He stated clearly that the order for this had come from the highest quarters, but that it should be done in his – that is Riccardi's – name. In the interim no further copies of the *Dialogue* were to be despatched from Florence to other places.

His letter contained a strange P.S. On the title page of Galileo's book was a sort of seal, with a drawing of three fishes, or possibly dolphins, swimming after each other. Riccardi insisted on being told without delay what these meant. Could they be the printer's seal?

The question had in fact come from Urban VIII. For some reason he had taken it into his head that the fish were a reference to his three nephews, of whom he had taken more than generous care – a fact which, of course, was general knowledge.

A reassuring message was quickly received from Florence to the effect that it *was* the printer's common seal. But the matter was serious all the same, for it indicated that the Pope had begun to incline to the idea that the *Dialogue* was a kind of treachery on Galileo's part, a deceitful attack upon himself.

This was partly connected with the dismissal of Ciampoli. The feeling of conspiracy on every side made Urban instantly perceive a link between him and Galileo, Lynceans as they both were. Indeed, on one occasion he even called the publication of the *Dialogue* a *ciampolata*[77] – a word of his own coining that meant something like "a nasty trick typical of Ciampoli".

Beneath this mask of injury, a more calculating scheme was hatching. If Urban were now to deal decisively in the matter of his former favourite

Galileo, he would be able to demonstrate two things at once. Firstly, that he really did take the orthodox faith and doctrine seriously, and secondly that he did not bestow unfair advantages on those who were close to him.

Just what part Father Scheiner and others close to the Collegio Romano played, has never been completely revealed. Galileo's friends in Rome were in no doubt. They thought it certain that "the Jesuit Fathers are working most valiantly in an underhand way to get the work prohibited", and the censor himself, the Dominican Father Riccardi, was reported to have said "The Jesuits will persecute him most bitterly"[78].

Much indicates that someone from this circle may have pointed out the concluding sequence of the *Dialogue* to Urban VIII, where his well known tenet about God's omnipotence was trotted out by *il semplicione*, the simpleton. Was this not an infamous trick, a neat way of inferring that the *Pope* was unsophisticated, that Urban VIII was the jester in the drama being played out between the Aristotelian-Ptolemaic system which Galileo so obstinately and obviously despised, and the Copernican, which the Church itself had clearly and expressly forbidden?

Riccardi at first understood that the Pope considered it necessary to make certain amendments and additions to Galileo's text. The formal grounds for this were that the book was not printed exactly in accordance with the manuscript that the censor had approved. But what actually had been approved in the confusing process prior to publication was not so easy to ascertain.

In early August Galileo received news from Rome about the hiatus in printing and distribution. He was furious, but did not lose courage. There must be an amicable solution to the problem if the Grand Duke interceded. After all the *Dialogue* was dedicated to him. He therefore applied to the court at Florence, which made contact with its Ambassador in Rome. Ambassador Niccolini lodged an official protest with Father Riccardi over the attempts to confiscate a book that had been lawfully printed in Florence, with the *imprimatur* of the local Inquisitor.

The Ambassador got an immediate reply, a reply which showed that in the course of a couple of hot summer weeks in Rome, the matter had taken a completely new turn. Now, he reported back to Florence, there was no longer any talk of small additions and corrections:

> "... I hear that there has been set up a commission of persons versed in his profession, all unfriendly to Galileo, responsible to the Lord Cardinal Barberini..."[79]

Tommaso Campanella had also heard about this commission. He was rather less diplomatic than the Ambassador when he wrote to Galileo:

> "I have heard (with great disgust) that they are having a commission of irate theologians to prohibit your *Dialogue*; and there is no one on it who understands mathematics or recondite things (...) I fear the violence of people who do not know. The Father Monster [Riccardi] makes fearful noises against it; and, says he, *ex ore Pontificis* [from the Pope's mouth]. But his holiness is not informed..."[80]

It was Campanella himself who was uninformed. He clung to his belief in a liberal Urban VIII for as long as possible, the man who had freed him from prison in Naples and given him a position in Rome. But the Barberini Pope was no longer the inquisitive and open intellectual.

Campanella's advice to Galileo was that Grand Duke Ferdinando had to intervene and demand that the commission be enlarged by two members, namely Father Castelli, Galileo's faithful pupil from Padua, who was now a professor in Rome – and Campanella himself!

The latter would have been of little help. The brave, colourful and highly unorthodox Campanella was rapidly falling from grace, greatly helped by someone who had dug up his prohibited work *In Defence of Galileo* from obscurity. Other things piled up and, two years later, in 1634, he found himself in deep trouble. After 27 years in prison he did not want further confrontations with the courts, and he fled from Rome in disguise.

The diplomatic post now began to fly back and forth between Florence and Rome. The Grand Duke's Secretary of State, Andrea Cioli and the Ambassador maintained that Galileo's case was legally unassailable: the *Dialogue* had been approved in accordance with the procedure that Father Riccardi had finally dictated. Therefore, no commission was required but, if one were to be appointed, it had to include representatives that were well disposed towards Galileo. They wisely refrained from putting forward names.

The Grand Duke's Ambassador did not approach the Pope over the matter for the time being. He contacted his nephew, Cardinal Francesco Barberini, who was to be directly responsible for the commission. Francesco spoke warmly and at length about the "goodwill" he bore Galileo, and gave assurances that the Pope himself still regarded the mathematician as a much loved and favourite friend. However, he made no promises to intervene.

The Ambassador in Rome got a more concrete idea of the nature of the problems from another source. They revolved around two points in particular. One was obviously that Urban's argument had been put in the mouth of Simplicio. The other concerned the preface. This was clearly separate from the rest of the book. It was placed before the first "day" and was also set in a different type. The preface might therefore give the impression of being "added on" – which, of course, it very much was.

But in spite of everything this reassured the Ambassador and his superiors in Florence. The objections were not so serious that they could not be dealt with by changes to the text, unless the commission came to quite a different conclusion. The Ambassador was to have an audience with Urban VIII about another delicate matter – a man who had been charged by the Holy Office, but whom Grand Duke Ferdinando did not want to hand over to Rome straight away.

The way matters now stood, it might be as well to take up the *Dialogue* with His Holiness directly.

Diplomacy in the Time of the Plague

The author of all these diplomatic, theological and legal complications had now rented a smaller and cheaper house a little further away from the centre of Florence, in Pian de' Guillari near Arcetri. The house cost 35 scudi per annum, and had the beautiful name *Villa di Gioiello* – "the jewel" or the "the gem". However, the most important thing for the ageing Galileo was to get closer to his daughters in the convent of San Matteo; from his new house it was only a few minutes walk away.

At about the same time, the war that still raged in northern Europe took a decisive new turn. In the summer of 1630, Gustav Adolf landed in Pomerania with a small force, ostensibly to secure the Lutheran position in northern Germany, but in reality to protect Swedish interests around the Baltic. At the Imperial court people barely raised an eyebrow, merely registering the fact that "another small enemy" had arrived. Perhaps the Emperor's counsellors were recalling Christian IV of Denmark-Norway and his attempt to intervene in affairs a few years earlier, which quickly ended when the imperial army leader Wallenstein chased the Danish king back to Copenhagen.

But Gustav Adolf was made of sterner stuff. It was true that the Protestant petty princes regarded him with scepticism. They feared – quite rightly – that his interests chiefly lay in dominating them. In the meantime he soon found a powerful ally with large financial resources. France's strong man might well bear the title of cardinal, but he felt no scruples about allying himself with a Protestant when the power of France was at stake. In 1631

France and Sweden signed an agreement. Almost simultaneously, Catholic troops took Magdeburg, one of the Protestants' strongest cities.

Even in a war that was remarkable for sustained brutality on all sides, the conquest of Magdeburg was a chapter that stood out. The city had 36,000 inhabitants. Only 6,000 of them survived the battles that were followed by out-and-out massacres. If Protestants had not worked together before, they now realised that it was imperative.

With support from north German allies and mercenaries in the pay of the French, the Swedish troops swept down through Germany, taking city after city, and quickly approached Vienna and Prague. In regions of Europe where the people had previously feared the Turks more than any other living thing, they now learnt that it was little better when the new cry went up: "The Swedes are coming!"

This was the dramatic development north of the Alps that had caused problems for Pope Urban and had been the backdrop to the scandalous consistory meeting in March. But in Tuscany and Florence the war was far away. Grand Duke Ferdinando II had just reached his majority and, fortunately for his subordinates, Tuscany's ruler no longer had any role to play on the European stage, where power and religion were becoming enmeshed in such an unhappy manner. He did, however, engage himself in local conflicts with the men of the church. The cause of this was his genuine attempt to do something about the most imminent threat to Tuscany: the plague.

Ferdinando, amiable and weak as he was, had, despite the orthodox rearing received from his mother and grandmother, a few shreds of Tuscan rationalism in his nature. One thing was the way he displayed personal courage by remaining amongst the plague-smitten, rather than fleeing to the countryside, but more important were the decrees he made and the bureaucratic apparatus he built up to limit its ravages.

The work was based on vague contemporary ideas of infection. They kept the sick isolated, limited social contact between people and exterminated the sources of infection. This opened the way for a conflict between "faith" and "science" even at the local level. Churchmen in villages believed that the best thing to do was to bring the local image of the Virgin Mary out of the church and organise a great procession for the villagers and the people of the surrounding countryside. The publicly appointed "plague officers" regarded this as a dangerous spread of infection. They in their turn suggested initiatives such as killing the village dogs and cats, which were suspected of spreading the disease in some way.

Such conflicts had no clear moral victor. The religious processions did not really do much harm, as the plague rarely spreads directly between people. Slaughtering cats and dogs, on the other hand, was decidedly unfortunate. It led to a boom in the rat population, which in its turn enabled the real culprit, the bacteria *yersina pestis*, to spread through its host, the rat flea.

Even though he had enough problems at home and with the Pope, and was not endowed with any particularly incisive ability to cut to the quick and solve them, young Ferdinando was genuinely interested in Galileo's fate. The mathematician had had a close relationship with his family for decades, and had brought glory and honour to Florence and the Medici family. It was therefore with the full support of the Grand Duke that his Ambassador in Rome went to the Vatican.

But it was a shocked and incredulous Ambassador Niccolini who returned to the Villa Medici after his audience on 4 September 1632.

As planned, he had commenced with the affair of the arrested Tuscan and his possible hand-over to the Holy Office.[81] Something was clearly bothering the Pope, however, and the Ambassador soon discovered what it was. Suddenly Urban VIII exploded in a violent fit of rage. *Galileo* too, he ejaculated, had gone too far and entered territory that was nothing to do with him, and had meddled in the most dangerous matters imaginable.

The Ambassador was no coward. And anyway, he had instructions to raise the matter of Galileo. As the Pope had now seen fit to introduce the subject, he felt he might as well continue. He therefore remarked that Galileo had not allowed the book to be printed without prior approval from the Pope's own men, and the Ambassador had himself helped the process by sending the drafts of the preface back and forth between Rome and Florence so that everything might be done properly.

This was undoubtedly true, but probably the worst thing he could have said just at that moment. For "the Pope's own men" included Ciampoli, who was still residing in Rome, but at a safe distance from the Pope's displeasure. In a further paroxysm of anger, Urban shouted that he had been tricked both by Galileo and Ciampoli, as the latter had intimated that everything was all right and that Galileo would do exactly as the Pope had ordered. For good measure he also blackguarded Father Riccardi, the censor who, beguiled by "fine words" had been wheedled into giving an approval, an approval which was subsequently exploited in Florence and even printed at the front of the book, even though it was only valid in Rome.

The Ambassador now saw the seriousness of the situation and the extent of the Pope's wrath. He hastily put in that he at least hoped that Galileo would be called to put his own explanation to the commission which rumour had it was to be appointed. But Urban did not give way. He replied tersely that the Holy Office did not work like that. A piece of writing had a judgement passed on it, and then the sinner was summoned so that he could, if necessary, renounce his opinions.

Of course, said the Ambassador. Even so, would it not be more practical if Galileo knew what was wrong in advance, what was troubling the Inquisition in this way?

This rash attempt at gainsaying him caused Urban to explode for the third time: the Holy Office did not do things in that way, it did not work like that, information was never provided in advance, it had never been done before. And anyway, Galileo knew perfectly well what was wrong: "We have discussed them [the objections] with him and he has heard them from ourselves."[82]

After this outburst in the papal plural, Urban calmed down a little. He went on to say that he did not care if the *Dialogue* was dedicated to the Grand Duke. In his role as the Pope he had personally banned books that were dedicated to him, and if Ferdinando wanted to be seen as a Christian prince, he should help by getting ungodly texts prohibited, not defending them. And anyway, he added in a slightly milder tone, he had already done everything he could for Galileo by appointing this special commission of pious and learned men instead of sending the *Dialogue* through the normal channels straight to the Holy Office. In brief, *he* – Urban – had been as accommodating as it was possible to be under the circumstances, while the opposite had be said of Galileo: the mathematician had done his best to trick and deceive his former pontifical benefactor.

With this powerful salvo the Ambassador was dismissed. It was not until the following day that he had sufficiently recovered to send the Florentine court a detailed report, which concluded:

> "Thus I had an unpleasant meeting, and I feel the Pope could not have a worse disposition toward our poor Mr. Galilei. Your Most Illustrious Lordship can imagine in what condition I returned home yesterday morning."[83]

As things now stood it was useless to try to influence the composition of the commission. Galileo, isolated in his Villa di Gioiello, far from the centre of events, still hoped for something of the kind. For the first and only time he listened to advice from Tommaso Campanella, and enquired of the Ambassador if it were possible to get Campanella or other sympathetic people on to the commission. For understandable reasons the Ambassador had little enthusiasm left for seeking out the Pope with such an enquiry, so instead he aired it with the censor, Riccardi, who was himself to be one of its members.

Riccardi replied truthfully that it would be quite impossible to have Campanella on an official commission of this kind. It had only been a few years since a book of his own had been on the Index, a book that dealt precisely with the relationship between astronomy and religion – *In Defence of Galileo*. And in any case, the censor added – possibly not quite so honestly – two members sympathetic to Galileo had already been appointed. One was himself, as he naturally had an interest in defending his own decision to approve the printing of the *Dialogue*, the other was the astronomer Melchior Inchofer. Father Inchofer was known to be a defender of the geocentric, Ptolemaic system, but he was a professional and would be able to assess Galileo's proofs and arguments.

This assurance was probably meant to do little more than hearten the Ambassador. The turn the situation had now taken meant that Riccardi's principal objective was to save his own skin. When he had to explain how the book had actually come to be printed, it would decidedly be easiest to argue that Galileo had pulled the wool over his eyes, especially as this was what the Pope wanted to hear. And as far as Father Inchofer was concerned, he was indeed a Jesuit, but no astronomer of any standing. In such matters he tended to defer to an older and considerably more skilful colleague: Father Christopher Scheiner.

The commission was fast working in the extreme. It had five meetings in the space of just a few days. Its conclusion surprised nobody: the *Dialogue* must immediately be sent to the Holy Office for thorough investigation.

The Grand Duke's sorely tried Ambassador steeled himself for another meeting with the Pope. It was a much more relaxed Urban VIII who met him this time, friendly and almost inclined to joke. The Pope assured him of his deep respect for the Grand Duke and that he still looked on Galileo as a friend. But about the continued handling of the case, he was immovable: the Inquisition would decide the future fate of the *Dialogue* and its author.

Perhaps Urban's good humour was just a chance expression of the extraordinarily labile shifts in mood which pressure and adversity had clearly brought out in him. But perhaps his humour was just slightly improved by a decisive document he was able to show the Ambassador, a sensational find from the archives of the Inquisition that placed all of Galileo's great labour over his *Dialogue* in a new and considerably more dubious light. The

Ambassador was to greet Grand Duke Ferdinando, said the Pope, and tell him that "the matter is more serious than His Highness thinks."[84]

An Order from the Top

No one knows who searched the archives. But one way or another it was Cardinal Robert Bellarmine who, twelve years after his death and sixteen after their last meeting, had yet again cast a shadow on Galileo's life.

In 1616 there had been strong rumours in Rome that Bellarmine had forced Galileo into a formal renunciation of his belief in the Copernican system. Certainly Galileo got the Cardinal to disavow the fact, but that had been done in a private statement that was not made public.

The archives of the Holy Office were not open to all and sundry. The members of the commission who assessed the *Dialogue* had no access to it, for example. In any case the whole idea of the commission was that it was to form an opinion of the book *before* the Inquisition – if required – became involved. Therefore neither Inchofer, nor Scheiner in his shadow, could have been responsible for the archive discovery.

It might, of course, have been an eager official going through the archive with a view to obtaining the best possible foundation for the case against Galileo. But the fact is that the rumours about this mysterious document began to spread before the matter was referred from the ad hoc commission to the Inquisition itself.

There is considerable evidence that the archive search was instituted a lot earlier, by someone who remembered the rumours of Bellarmine's intervention. If a formal document existed, one in which Galileo promised to keep away from the ideas of Copernicus, it would obviously place him in an extremely difficult situation now that he had conspicuously written a book that thoroughly aired the self-same ideas.

Pope Urban VIII – at that time still Maffeo Barberini – was himself in Rome in 1616 and took part in the process as a member of the Congregation. Of course he remembered the talk, even if he did not necessarily believe it. Even if he was not responsible for finding the document, he certainly did nothing to lessen its effect or prevent its being used.

For this sensational discovery was a somewhat dubious document. Naturally it did not come from the hand of Bellarmine, as the Jesuit cardinal had given Galileo a friendly, if unambiguous, warning against presenting Copernicus' ideas as a description of physical reality. The document was not signed, and therefore of highly debatable legal value.

But there was little doubt that the discovery in the archives, if it was accepted in evidence and taken at face value, presaged even greater hardships for Galileo. The document was in fact Cardinal *Segizzi's* version of the meeting in the Paradise Rooms, Bellarmine's residence, on 26 February 1616. It was pretty well a word for word rendering of the salvo Galileo had got from Segizzi, after Bellarmine's moderate warning had seemingly not sunk in. There it stood clearly in black and white:

> "The said Galileo was (...) to relinquish altogether the said opinion that the Sun is the centre of the world and immovable and that the Earth moves; *nor further to hold, teach, or defend it in any way whatsoever*, verbally or in writing; otherwise proceedings would be taken against him by the Holy Office; which injunction the said Galileo acquiesced in and promised to obey."[85]

Nec quovis modo teneat, doceat aut defendat.

If forced, Galileo might possibly maintain that he had not "in any way whatsoever" held to Copernicanism, merely presented it. But no reader of the *Dialogue* would be in any doubt that he had "taught" many fine points that sprang from the heliocentric system and, as for "defending" it, the Salviati character did almost nothing else during its 500 pages.

And what was even worse: if this document were to form the basis, it would not help Galileo one jot to say that he sincerely believed that both the censors and Pope himself had acceded to a "discussion" of the sort he had committed to paper, and that he had conducted an ongoing dialogue with Urban VIII about the problem for years. Because it would then be clear that he should never have been concerning himself with the subject at all!

On 23 September 1632 the Inquisition assembled to begin the process against Galileo. The Pope was present in person, together with eight of the ten cardinals who were the heads of the Holy Office. During the meeting a report was submitted about the unusual circumstances surrounding the approval and printing of the *Dialogue*, so too was an opinion from the commission which had gone through the book.

The document from 1616 was also presented, without objections. Or rather, the minutes of the Inquisition's meetings do not describe disagreements or dissent, but rumours in Rome immediately afterwards had it that one of the cardinals courageously stood up for Galileo and moved that the matter be dropped. If so, he was heavily out-voted. The meeting ended with Urban VIII giving orders that a letter be sent to the Inquisitor in Florence. He was to visit Galileo at home together with a notary and some witnesses, and convey an order to him: Galileo must present himself at the Holy Office before the end of October.

At the Villa di Gioiello Galileo waited, sleepless and rheumatic, for the Ambassador's work in Rome to yield results. In the meantime he pottered about with the grape harvest and wine making – this house, too, had some farmland, with vines and fruit trees. His old optimism had not deserted him, he hoped that the prohibition of the *Dialogue* would be rescinded, or at least that he would be instructed about alterations to the text.

Instead he received an unexpected visit. The Holy Office did not waste time once a case was under way. On 1 October the local Inquisitor made his way up to the village from Florence, in company with a notary. He had the order from Rome.

In its formal language the summons to attend the Holy Office in person was read out to Galileo. The old man acknowledged before the Inquisitor and his retinue that he had understood the order and would obey.

Behind his facade he was stunned. Until the very moment he had heard the words the Inquisitor read out, he fully believed that the whole thing was about his *book*. That was unpleasant enough. But this was something quite different. Now, suddenly, there was talk of his own, frail and aged, *person* – it was nothing to do with a work on the *Index*, but a charge at the court of the Inquisition.

There they were not concerned with deviant views of a greater or lesser kind, which might ultimately be adjusted and corrected. The Inquisition dealt with only one crime: heresy.

Galileo was not completely alone with his worries. He had a housekeeper and a servant boy living in his house. He could visit his son Vincenzio, with whom he was now on good terms, and he could walk the short distance to the convent and talk to his wise elder daughter. But none of these could give him advice in this critical situation. The Grand Duke and his court were at Siena, and naturally he wrote there at once. Even so, it was pretty clear that Tuscany had *already* given him all the official help it was able to, without it having done the least good.

If he were to find anyone to help, it would have to be in Rome. He decided on his old friend, the Pope's nephew, Cardinal Francesco Barberini, who occupied a top position in the Holy Office. Galileo wrote to the Cardinal:

> "While I go on pondering to myself the fruits of all my studies (...) those fruits are turned into serious accusations against my reputation by encouraging my enemies to rise up against my friends and sealing their voices not only as to praising me but also as to excusing me, with the allegation

> that I have finally merited to be cited by the Tribunal of the Holy Office, an action which is not taken except in the case of those who are seriously delinquent."[86]

This was having such an effect on him, he wrote, that he was unable to sleep. And he listed his many physical ailments. He went on to suggest two possible ways of settling the matter: he could write a detailed account of all his work on the ideas of Copernicus and send it to the Holy Office. Galileo – still not entirely bereft of optimism! – thought this ought to be enough to show that he was innocent.

If a written document was not good enough, his alternative suggestion was that he could make a deposition to one of the ecclesiastical dignitaries in Florence: the Inquisitor, the Papal Nuncio, the Archbishop. He would do everything possible to accommodate such an arrangement.

Naturally Galileo was aware that one did not negotiate with the Inquisition – one simply submitted to it. He therefore rounded off his letter to Cardinal Francesco with the following peroration, which at least showed that he had lost none of his skill with words:

> "And finally to conclude, when neither my advanced age nor my many bodily ills, nor my troubled mind, nor the length of a journey made most painful by the current suspicions, are judged by this sacred and high Court to be sufficient excuses for seeking some dispensation or postponement, I will take up the journey, preferring obedience to life itself."[87]

The letter did no good at all.

The Grand Duke tried as well, with a direct and respectful application to Urban VIII, pointing out Galileo's advanced age. Indefatigable Ambassador Niccolini was mobilised yet again and got an audience. The Pope was as unyielding as before: hopefully God would forgive Galileo, he said, for becoming involved in such an intrigue after he, His Holiness, when a cardinal, had saved him from it.

Precisely what Urban meant by this is impossible to say, but there can be no doubt that he felt injured and offended.

Galileo's deadline was about to expire. October drew to a close, but he did not set off. When the Inquisitor again visited, he said that he wanted to go, but that he was prevented by illness. The Florentine Inquisitor could see that he really was unwell and, on his own initiative, allowed him another month, writing to Rome at the same time: "... and he showed himself ready to come; but then I do not know whether he will carry it out."[88]

The Holy Office grudgingly approved this deadline. But a message was sent back to Florence saying that when it had expired, Galileo *was* to set out,

Grand Duchess Christina: pious, powerful and imposing. (© akg-images)

The bronze manifestation of pontifical power. Bernini's baldachin in St. Peter's (1633). (© Corbis)

Baroque elegance, but little of the old Medicean power: Cosimo II with Grand Duchess Maria Maddalena and their son, later to become Ferdinando II. Painting by Justus Sustermans. (© Corbis)

DIALOGO

DI

GALILEO GALILEI LINCEO

MATEMATICO SOPRAORDINARIO

DELLO STVDIO DI PISA.

E Filosofo, e Matematico primario del

SERENISSIMO

GR.DVCA DI TOSCANA.

Doue ne i congressi di quattro giornate si discorre sopra i due

MASSIMI SISTEMI DEL MONDO

TOLEMAICO, E COPERNICANO,

Proponendo indeterminatamente le ragioni Filosofiche, e Naturali tanto per l'vna, quanto per l'altra parte.

CON PRI

VILEGI.

IN FIORENZA, Per Gio:Batista Landini MDCXXXII.

CON LICENZA DE' SVPERIORI.

Nr. 293

The title page of the *Dialogue*, published in Florence in 1632. "Advancing without conclusion the philosophical and natural arguments, equally for the one part as for the other" (the text above the seal with the fishes). (© akg-images)

Sister Maria Celeste, Galileo's daughter. Her religious name was probably chosen by her to reflect Galileo's work: Celeste means "heavenly". (© Welcome Library, London)

This was the heroic 19th century depiction of the collaboration between the aged Galileo and Vincenzio Viviani, his last disciple. (Tito Lessi). (© Photographic Archive, Institute and Museum of the History of Science – Florence)

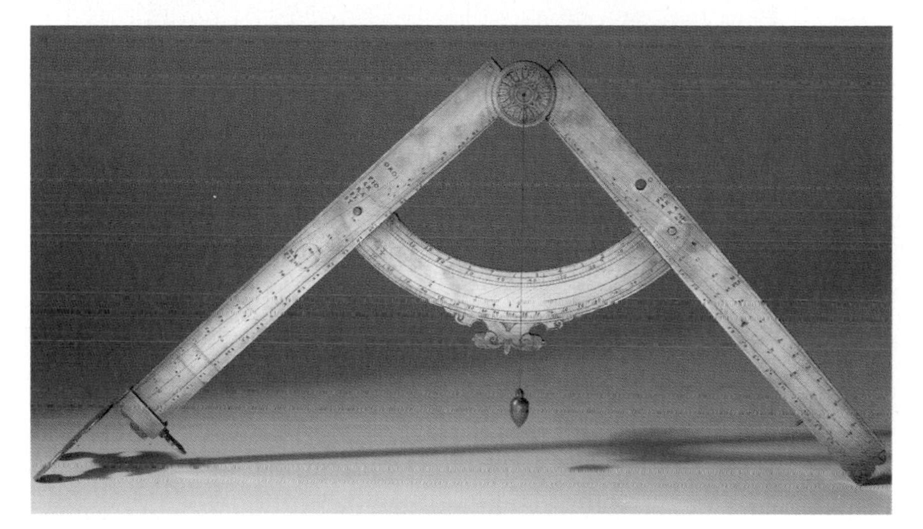

Galileo's compass. (© Photographic Archive, Institute and Museum of the History of Science – Florence)

Galileo before the Inquisition. A coloured lithograph of 1865, from the drawing by Albert Chereau. (© akg-images)

Ptolemy's notion of the world from a copperplate engraving circa 1500. The ring of fire is not the Sun, but indicates the earthly elements' place below the Moon. (© akg-images)

Padua – Galileo's observation tower. (© akg-images)

Venice, a busy seafaring city. In the background is the Campanile (Rudolf von Alt). (© akg-images)

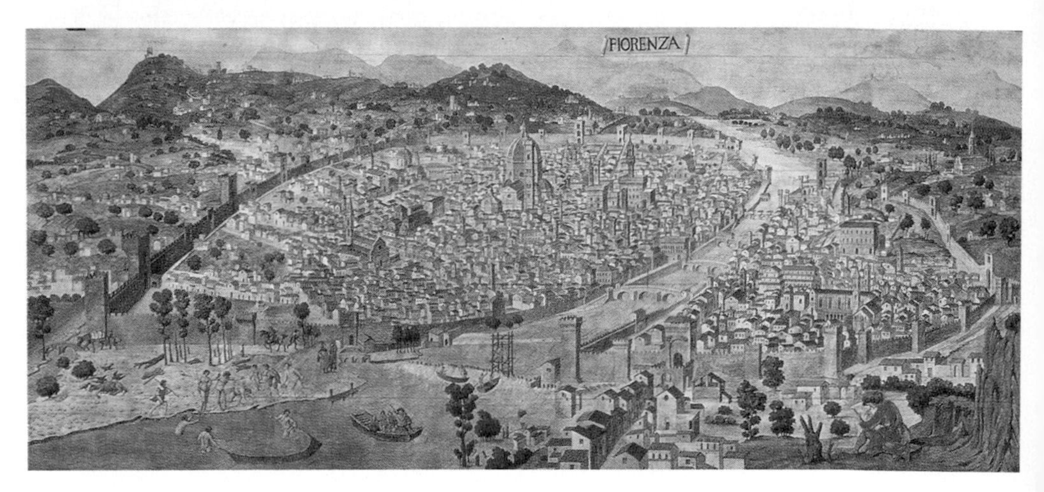

The Florence of the Medicis. Copperplate engraving from around 1490 (Francesco Rosselli). (© akg-images)

no matter what. His friends in Rome realised that Galileo's hesitation might be used as yet another indictment against him – a sort of "contempt of court" – and urged him to try to begin the journey.

There is no doubt that Galileo did his best to get out of going. But his illness was real enough. In one final attempt he summoned three doctors and got them to write a certificate. This was too much for Urban who stated that: "His Holiness and the Sacred Congregation cannot and absolutely must not tolerate subterfuges of this sort."[89]

This was followed by the definitive order: if Galileo would not come voluntarily, he would be brought to Rome in chains.

On 15 January the old man sat down to write his will. It was brief, he bequeathed most of his goods to his son Vincenzio. After that he was ready for his final journey to Rome.

"Nor Further to Hold, Teach, or Defend It in Any Way Whatsoever"

After his long and laborious journey Galileo stayed with the Tuscan Ambassador. Formally he was not a prisoner in the Villa Medici. It was just a "friendly piece of advice" from the Holy Office that he should not leave the property.

It fell to the Ambassador to gauge the mood and find out whether there were still channels of influence open. He soon realised that the worst problem would be the document from 1616 containing the unfortunate words: *Nec quovis modo teneat, doceat aut defendat.* But when he gently hinted this to Galileo, the old man reacted with agitation and confusion. He certainly could not recall being given any such order. He had been summoned to Bellarmine and been given a warning not to present Copernicanism as a physical reality, but that was quite a different matter!

Pope Urban VIII was less choleric than the previous year, but just as unyielding regarding the case. He emphasised how leniently Galileo had been treated, especially in living with the Ambassador instead of being thrown into the Inquisition's gaol. But he could not promise any speedy resolution: "... the activities of the Holy Office ordinarily proceeded slowly,"[90] the Ambassador reported him as saying. Besides, he was still lambasting Galileo for working with the arch-villain Ciampoli.

The Ambassador was pessimistic, although he did not show his feelings to Galileo. But he wrote to Grand Duke Ferdinando:

> "... even if they should be satisfied with his answers, they will not want to give the appearance of having made a blunder, after everybody knows they summoned him to Rome."[91]

He also had a sense of just how virulent the antipathy, even the hate, towards Galileo was in some quarters – most probably especially amongst the Jesuits close to Grassi and Scheiner.

This tremendous aggression was noted by another observer, the German Catholic, Lukas Holstein. He was an outsider and saw the situation with fresh eyes. He was worried about the real problem as well, which in Rome had been completely obscured in the excitement surrounding declarations and formulations: what would happen to the Church's authority if Copernicus was right after all?

> "It would take a long time to report the cause of the hatred harboured against the very fine old man [Galileo] but one thing cannot be seen without irritation, that is, that persons completely incapable have been given the task of examining the book of Galileo and the whole Pythagorean and Copernican system, while it is above all a matter of the authority of the Church which will suffer widely from a less correct judgment. Galileo suffers from the envy of those who see in him the only obstacle to their having the reputation of the highest mathematicians. Because this whole storm was raised by the personal hatred of a monk whom Galileo does not wish to recognise as the first among mathematicians (...)"[92]

It was the Ambassador who received the news that the hearing was imminent. In a final attempt he visited the Pope yet again, on the pretext that he was thanking him, on behalf of the Grand Duke, for the special treatment Galileo had been promised – he was not to be incarcerated in a cell, but live in an ordinary room under light guard. The Pope was calm but inflexible:

> "His Holiness complained that he [Galileo] has entered into that matter which for him [the Pope] it is still a most serious matter and one that has great consequences for religion."[93]

It was the Ambassador himself who had the unpleasant task of telling Galileo about the trial.

The old man took it very hard. Fears, sleeplessness and rheumatic pains prostrated him to such an extent that the Ambassador feared for his life. But no postponement was mentioned. The Ambassador earnestly advised him not to try to defend himself, but to submit to any objection the judges of the Inquisition might raise, and have faith that out of his own fame, and political deference to the Grand Duke, a lenient sentence would result.

On 12 April 1633 Galileo was taken from the Villa Medici, through the streets of Rome, across the Tiber, to the headquarters of the Holy Office. There he was held as a prisoner. But he was lodged in rooms that were intended for the use of officials, and he was allowed to go out into the courtyard. The servant who had come with him from Florence was permitted to attend him, and the embassy servants could bring him food twice a day.

Once the interrogation began, however, the tone was entirely formal. Present were the Commissary of the Inquisition, Father Maculano, together with witnesses and a notary. The other cardinals were content – as usual – to read the summary and form their opinions based on that.

The interrogation began with the usual questions about name and background – and about how much Galileo knew about why he had been summoned. He answered deferentially that he assumed it must have something to do with "my book which has just been printed", and which he gave a short resumé of. He was then shown a copy of the *Dialogue*, and confirmed that he had written it and was responsible for everything it contained. In reply to a question about how long he had taken to write the work, he answered that he had begun ten or twelve years ago, and spent perhaps six or eight years on it with breaks in between.

This was mere formality and preliminary skirmishing. Instead of going further into the book and its contents, the Commissary suddenly changed the subject and asked Galileo if he had been in Rome previously, and particularly in 1616.

But Galileo was prepared. He answered calmly that he had travelled to Rome on his own initiative in 1616 – and that, furthermore, he had been in the city twice afterwards, "in the second year of His Holiness Urban VIII's pontificate", and in 1630 to organise the printing of his book. And so, without saying it openly, he managed to emphasise that his work on the *Dialogue* quite literally had continued with the blessings of those in the very highest places.

Father Maculano had no interest in listening to Galileo's connections to Urban and the papal court. He turned again to 1616 and what had happened then. Why precisely had Galileo come to Rome?

The old man replied that some of the cardinals, including Bellarmine, wanted to have an explanation of Copernican theories, which were extremely hard for laymen to understand.

And what emerged from these discussions and explanations? asked the Commissary.

Galileo had to admit that it resulted in a statement from the "Holy Congregation of the Index" saying that Copernicus' doctrine contradicted

Holy Scripture if taken literally and that it was only to be used hypothetically (*ex suppositione*) – just as Copernicus had done, he added piously.

This last was in fact an evasion of the truth, but the Commissary did not pick up on it. Instead he followed the obviously pre-arranged plan, and asked how and from whom Galileo had heard of this decision.

This was serious stuff. Galileo immediately admitted that he had been personally informed of it by Cardinal Bellarmine. But he insisted that Bellarmine had expressly said that "Copernicus' theory could be presented *ex suppositione*, just as Copernicus himself had presented it".

Galileo obviously felt fairly secure. He held a trump card in his hand and now he played it: he submitted a letter to the court, the certificate that Bellarmine himself had written in May 1616, just before Galileo had returned to Florence. It explained that Galileo had simply been informed of the decisions of the Inquisition and Congregation of the Index, and that there was no question of refutation or punishment.

Commissary Maculano now had two contradictory documents before him: Bellarmine's sober statement and the severe, unsigned document which had originated from Cardinal Segizzi and about which Galileo had as yet not been properly informed. Maculano now went to the heart of the matter via a tactical diversion:

Were there any *others* there with Bellarmine on the day Galileo had been warned not to take Copernicus literally?

Yes, said Galileo. There had been some Dominican Fathers there, but he could not recall their names, nor had he met them subsequently.

Now Father Maculano stuck the knife in: had any prohibition (*praeceptum*) been issued on that occasion, by the Dominicans or any others?

Galileo's answer was strange:

> "I remember that the transaction took place as follows: the Lord Cardinal Bellarmine sent for me one morning and told me certain particulars which I had rather reserve for the ear of His Holiness before I communicate them to others. But the end of it was that he told me that the Copernican opinion, being contradictory to Holy Scripture must not be held or defended. It has escaped my memory whether those Dominican Fathers were present before or whether they came afterward; neither do I remember whether they were present when the Lord Cardinal told me the said opinion was not to be held. It may be that a command [*precetto*] was issued to me that I should not hold or defend the opinion in question, but I do not remember it, for it is several years ago."[94]

For the last time Galileo here attempts to exploit the special ties of friendship he had to Pope Urban VIII Barberini. It is impossible to say what information from Bellarmine he wanted to convey to Urban. Presumably, Bellarmine said something to the effect that the then Cardinal Barberini looked with favour on Galileo's work – something everyone knew at the time anyway.

But Commissary Maculano pretended that he had not heard. Not a single word does he say about the "certain particulars" that Galileo will not divulge to the Inquisition. He knew that Pope Urban's former closeness to Galileo is a subject that must not be mentioned in this context, and certainly not by the defendant himself; it could do nothing but embarrass the Pope.

Instead the Commissary followed his plan of attack: could the defendant not recall a promise not to "hold, teach or defend in any way whatsoever" the Copernican doctrine – and who exacted it?

Wisely, Galileo refrained from denying that such a thing could have been mentioned. But if it was, he did not remember it, because he acted upon Bellarmine's written resumé, and it said nothing about "in any way whatsoever" or "teaching".

Even so Maculano asked: how could he consider writing the *Dialogue*? Had he got special permission?

No, replied Galileo, nor did he require it. For the *Dialogue* in no way attempted to hold, teach or defend Copernicus' theory – on the contrary it tried to repudiate it!

This assertion must have struck Maculano as remarkable to say the least. Galileo had certainly been advised to be pliant, but this was going a bit far, especially as he was under oath. It is doubtful if the Commissary himself had read the book, but he did have the expert opinion from the committee constituted the previous autumn to go on.

He did not follow up this comment either, but turned instead to the circumstances surrounding the permission to print, and the interrogation ended soon after with Galileo still on the retreat: Copernicus' arguments were weak (*invalide*) and not conclusive. After this he signed the minutes of the interrogation, swore himself to secrecy about what had passed – the Inquisition's actions were so secret that not even a person accused or sentenced was allowed to say anything about them – and was conducted to his comfortable prison.

Convinced with Reasons

And there, in the Inquisition's rooms, Galileo remained for a good while. His intellectual somersault had given Maculano a problem: if Galileo were to be taken at his word, namely that the *Dialogue* was a fundamentally anti-

Copernican text, then the entire foundation of the indictment fell away. But if that was the case, practically everyone who had actually *read* the book, had taken it the wrong way.

The Commissary needed a precise opinion on this point. To save time – and probably to be sure of the result, too – he reconvened the committee which had read the *Dialogue* the previous autumn. From a legal point of view this procedure was presumably quite legitimate: their first opinion had been an informal evaluation which was not carried out at the behest of the Holy Office, but by order of the Pope himself. Now the three members were asked to reply to a simple question: had Galileo overstepped the prohibition against holding, teaching or defending in any way whatsoever, the theory that the Earth moved and the Sun stood still?

The answers came during the course of the next few days, and they were unanimous in all matters of substance. Galileo had certainly both *taught* and *defended* the Copernican theory, and he was strongly (*vehementer*) suspected of holding it as well. The Jesuit, Father Inchofer, delivered the longest and severest opinion, making certain to point out that amongst Galileo's sins were the attacks on Scheiner:

> "Galileo's most important aim this time is to attack Father Christopher Scheiner, who has recently written extensively against the Copernicans: but this is nothing less than defending, and disgracefully wishing to maintain, the doctrine of the Earth's motion..."[95]

In a strictly legal sense, Father Inchofer was completely right in his judgement. Galileo *had* held, taught and defended the heliocentric system and was guilty. The fact that it was not just the law – or for that matter theology – that counted within the Catholic Church, but the entire, convoluted matrix of connections, protectors and influence, he was shortly to experience himself. Inchofer had to quit Rome in disgrace after arguing against the practice of castrating young boys in order to keep their singing voices pure. This could hardly be called a particularly heretical viewpoint – but the choir in the Sistine Chapel needed castrati, and Inchofer was exiled to Milan!

The opinions of Inchofer and the others did not, however, solve Father Maculano's problems, but rather created new ones. In the first place they showed that Galileo had given a false explanation during his interrogation: the *Dialogue* could in no reasonable sense be seen as an attack on Copernicus. Secondly, in legal terms it would be far worse if they had to assume Galileo *held* the Copernican opinion, rather than that he had merely taught or defended it. The latter two could, at a pinch, be seen as irresponsible intellectual exercises on a hypothetical or theoretical level. But to *hold* a view that was expressly forbidden, both by the Inquisition and the Congregation of the Index, was a serious misdemeanour: it was heresy.

Maculano alone had the responsibility for the further conduct of the case. The Pope had retreated to Castelgandolfo with his nephew, Cardinal Francesco. Galileo waited, impatient and anxious. One week passed, then two without any word from the court. Maculano took the matter up at the weekly meeting of the heads of the Holy Office on 27 April. The cardinals agreed that Galileo had been dishonest in his statement, that he had plainly denied what anyone might read in the *Dialogue*. But they also agreed that the matter still posed "various difficulties".

These difficulties were not of a legal or theological character, and certainly not connected with natural science, but were linked to Galileo's position and reputation. Although it was important to set an example that showed that Urban was an orthodox and reliable Catholic, the Papal States could not afford to disregard the relationship with Tuscany and the Grand Duke entirely. A discreet solution was to be preferred, and Father Maculano believed that the Pope, too, had expressed a similar desire.

And so Commissary Maculano asked the Cardinals' permission to try some private conversation with Galileo, without witnesses or minutes, to get the defendant to perceive his true fault. In this way the next official interrogation could go without a hitch and lead to the result that everyone wanted: Galileo's unconditional admission and statement of his "intentions" – the sinful motives that had brought him into the path of heresy. This last was very important in reaching a judgement and sentence.

It was agreed that attempting such a conversation might be profitable, and so Maculano visited Galileo a few days later.

Two weeks of the admittedly benign "prison" had clearly made the proud and bellicose mathematician so tractable that he no longer insisted on farfetched readings of the *Dialogue* to get himself off the hook with his honour intact. But it is also very probable that Father Maculano in courteous and seemly terms reminded him of a well known aspect of the Inquisition's practice: "convincing with reasons".

Or, as it was sometimes also called – *esame rigoroso*, "rigorous examination".

There is no doubt that plain torture was a normal part of the Inquisition's working practice. The commonest form it took was the *strappado*, in which the victim's hands were tied behind his back and he was then raised by his wrists, sometimes with weights attached to his feet. A large assortment of alternatives were available to the Commissary – thumbscrews, "Spanish boots" and the much-feared water torture, in which water was poured into the mouth until the victim was on the point of suffocation.

Galileo knew about these "convincing reasons" – as did everyone else – no matter how secret the Inquisition's decisions and methods were supposed to be. There was therefore no reason to threaten him with torture directly, or to show him the instruments of it, which was also a part of normal procedure.

Father Maculano and his clerical colleagues in the Holy Office did not want to lay hands on Galileo if at all possible. They preferred to deal with written abstracts and not with forcing out information and admissions. Their prisoner was highly respected, and he was old and frail. The Inquisition's bureaucratic procedures included examining a prisoner before the use of torture to ascertain whether he or she was strong enough for it. This rheumatic sixty-nine-year-old with his host of other ailments would hardly pass such a test, if it was to have any meaning at all.

Galileo understood these signals, there is no doubt about that. He admitted his fault, was contrite and willing to formulate an admission to the court – indeed, he would sit down immediately and begin it. Three days after his conversation with Father Maculano, he again appeared for formal interrogation.

This time the session was a short one. The Commissary put only one question: had the defendant anything he wanted to say?

The defendant had. It had "occurred to him" to read the *Dialogue* again, something he claimed not to have done for three years. He wanted to see if, "despite his purest motives" certain formulations could have emanated from his pen that might be construed as contrary to the Church's ordinances. And alas, he was forced to admit, so it had proved. A reader who did not understand his real motives – which were to disprove Copernicus – might easily gain the impression that the very arguments Galileo was trying to refute, appeared the most compelling. This was particularly the case with the discussion of sunspots and tides, arguments which Galileo honestly and sincerely considered uncertain and unconvincing, but which unfortunately had been made to seem thoroughly incontestable.

As regards the main motives for his actions, he had to admit that they sprang, first and foremost from "vain ambition". It was a natural tendency in human beings, he said, to admire their own perspicacity and to want to appear more astute than their fellows, even though in this case it was a matter of promoting unsound theories. He quoted Cicero: *Avidor sim gloriae quam sat est*, "I am more keen for Glory than is merited". If he were to write the book again, he would have been more careful not to give these false arguments such convincing power.

After this admission the session was over, and Galileo signed the minutes and took the usual oath of silence. After that he was taken back to his rooms. But on the way he must have had second thoughts, for the records of the case relate that he quickly returned to the court chamber asking to add something.

To make it quite clear that he did not subscribe to the forbidden theory that the Earth moved, he had a suggestion to make. The *Dialogue* ended with Salviati, Sagredo and Simplicio agreeing to meet again and continue their discussions. Thus there would be no difficulty in adding another "day" or two. Here Galileo would revisit the arguments that had been advanced for the prohibited theory, "and to confute them in such most effectual manner as by the blessing of God may be supplied to me."[96] He ended by asking the court – "this holy Tribunal" – to let him have an opportunity of realising his plan.

It is hard to know just how Maculano reacted to this absurd suggestion. In one way it could, of course, be seen as proof that Galileo's rebelliousness and pride had been completely broken, and that he regretted it so much that he was now even willing to disavow his dearest work, the work that had cost him so many years' labour. But to anybody who knew Galileo's previous output and writing style, the idea *could* also be interpreted as a new link in his subtle strategy of promoting dubious ideas under a thin veneer of formal reservation. If Galileo was given permission to add several chapters, this prohibited book would come into print, and it would then be up to the *reader* to weigh the arguments – not the relevant ecclesiastical authority.

But Maculano was certainly not displeased at this turn of events. He gave sudden and surprising permission for Galileo to move back into the Tuscan Embassy in the Villa Medici. The Ambassador was astonished, but happy for Galileo, and he also got the clear impression that Maculano was now working with Cardinal Francesco Barberini to get the matter disposed of as discreetly as possible.

May had arrived in Rome, and Galileo looked on developments with renewed optimism. On 10 May he was summoned to the Holy Office once more, this time to submit his formal defence, which he was entitled to do under the ordinances.

It was a summary of events as Galileo himself had understood them. He had indeed been warned by Bellarmine, but the warning had only touched on presenting the Copernican system as a description of reality. No, he could not remember any direct orders with the fateful words *teach* or *in any way whatsoever*. If they had been said, he had forgotten them, all the more so since his recollection was guided by Bellarmine's written account. He had not mentioned the warnings from 1616 to the censor, Father Riccardi, for the simple reason that he believed he was doing nothing wrong in writing the *Dialogue*.

But he admitted his conceit and his desire to shine intellectually, and accepted that sections of the book were not well formulated and ought to be changed. He did not directly repeat his suggestion of a new edition with added chapters, limiting himself to assuring them that he would repair the damage "with all possible expediency" – in any way their holiest of eminences, the Cardinals, "commanded or permitted" him to. He concluded by detailing his poor state of health and begged to be treated with "indulgence and leniency".

Afterwards Galileo was allowed back to the Embassy. The Ambassador thought that the matter would now be resolved within the month. He realised, of course, that there was no hope for the *Dialogue*, and sent word to Florence that Galileo would presumably be sentenced to a symbolic punishment for having ignored Bellarmine's warning. Because the mathematician still clung to the hope that the book would be published in one form or another, he had not the heart to mention this directly to him.

Another good sign was that Galileo was given permission to leave the precincts of the Embassy for short walks. Maculano had also promised to come to the Embassy; the Ambassador assumed that this was to arrange the final details prior to the closure of the case.

But Father Maculano did not come. May passed without a word from the Holy Office for Galileo. The Ambassador grew anxious and used his contacts – finally going to the Pope, who had now returned from his sojourn at Castelgandolfo. What he heard made him even more uneasy.

Finally Galileo got a summons. On the morning of 21 June 1633 he was to attend a new interrogation.

"I, Galileo Galilei"

The case was not, in fact, as simple as Maculano and Francesco Barberini had hoped. The attempts to send Galileo home with a friendly warning and a symbolic punishment, a certain number of penitential prayers, for instance, met with resistance. Certain people were not satisfied with

Galileo's explanations to the court. We do not know who they were – it may have been Jesuits in the Inquisition or the Pope himself. At all events, the legal interrogations of Galileo were augmented by a detailed indictment, *Contro Galileo Galilei*, put together at the offices of the Inquisition.

This document commenced with an uncritical repetition of the old accusations from Florence, those that stemmed from the Dominican fathers, Lorini and Caccini . It was especially the latter's loose claims and rumours that helped put Galileo in a very poor light. In Caccini's completely distorted version, the objective account in *Letters on Sunspots* was turned into a wholly pro-Copernican treatise.

Next, the document moved on to Bellarmine's warning, but even here its rendition was imprecise – it mixed up Bellarmine's oral exhortation with the unsigned, written minute which must have come from Cardinal Segizzi. Bellarmine's written affidavit – Galileo's most important weapon – was, in contrast, swept aside in a couple of lines.

Put in this context, Galileo's work could be viewed as fifteen to twenty years of rebellious, and more or less heretical, activity. As for the praise Maffeo Barberini had given him in his time as a cardinal, or the encouragement he was still receiving during the Pope's first years on the Holy throne, not a word was mentioned.

Was this document to form the basis of the treatment of the case, or should one look to Galileo's explanation before Maculano, possibly taking into account his age, state of health and connection to the Grand Duke?

The Holy Office was in principle an independent assembly which came to its own conclusions. But it is obvious that in this particular case, in which Pope Urban VIII was heavily involved, the Pope's judgement would be decisive.

And the Pope was inflexible. Galileo had put forward a clearly heretical assertion, which "contravened the Holy Scripture dictated by the mouth of God", and must be imprisoned because his action had been directly contrary to the order of 1616.

And so, in reality, the cardinals had little choice.

During the meeting of the Holy Office on 16 June, the outlines of the final interrogation were planned. The document of indictment was produced, approved without dissent and given the following endorsement:

> "*Sanctissimus* [the Pope] *decrevit* [decreed] that the said Galileo is to be interrogated on his intention, even with the threat of torture, and, *si sustinuerit* [once having undergone this examination of intention], he is to abjure

> *de vehementi* [under vehement suspicion of heresy] in a plenary assembly of the Congregation of the Holy Office, then is to be condemned to imprisonment at the pleasure of the Holy Congregation, and ordered not to treat further, in whatever manner, either in words or writing, on the mobility of the Earth and the stability of the Sun; otherwise he will incur the penalties of relapse. The book entitled *Dialogue of Galileo Galilei the Lyncean* is to be prohibited."[97]

The Ambassador had learnt about most of this, but true to his custom he had kept the worst from Galileo, saying only that the *Dialogue* was likely to be banned. It was therefore a somewhat unprepared Galileo who appeared at the interrogation on 21 June.

Maculano first asked if the defendant had any more to say.

Galileo replied that he did not have anything of importance to add.

The Commissary then went straight to the heart of the matter. Did Galileo, now or previously (and in which case, when), hold that the Sun was the centre of the world and that the Earth was not, but was in motion, and also had a diurnal rotation?

Long ago, before the decision of the Congregation of the Index and before the warning, said Galileo, he had been neutral and had viewed both models, the Ptolemaic and the Copernican, as feasible, that one or the other might accord with reality. But after the decision, all doubts were gone, because he was convinced of the wisdom of the Church. Therefore he believed fully and unreservedly in Ptolemy's model: the Earth stood still and the Sun was in motion. The *Dialogue* was written to present the different possibilities and emphasise that truth must be found in "higher thought".

Maculano said that his book did not give that impression. There it appeared that Galileo still believed Copernicus, or had at least done so when he wrote it. Therefore – if he did not decide to tell the truth, the court must have recourse to the "appropriate remedies".

Perhaps it was only at this point, that the gravity of his situation hit the old man.

But now, with quiet dignity, he held to his own line. He was finished with making excuses by posing as a misunderstood anti-Copernican. He replied:

> "I do not hold and have not held this opinion of Copernicus since the command was intimated to me that I must abandon it; for the rest, I am here in your hands – do with me what you please."[98]

Father Maculano repeated his warning, and this time completely bereft of euphemisms: Galileo must speak the truth, *alias devenietur ad torturam* – or "they will otherwise have recourse to torture". Galileo answered:

> "I am here to submit (far l'obbedienza); and I have not held this opinion after the determination was made, as I have said."[99]

Here the interrogation ended. Galileo signed with trembling hand, and was sent to the "prisoner's room" in the Holy Office where he had resided before. He had not been allowed back to the Embassy.

There he sat, all that afternoon, evening and night. During the course of the long, lonely hours all his optimism evaporated – now it was merely a question of just how total his defeat would be. He had plenty of time to think through the unambiguous threat of torture: was it just a formal part of the legal process of the trial – or was there a real chance of him being taken down to the Inquisition's cellars at daybreak?

His defence was that he had not literally believed in the teachings of Copernicus since 1616. But would he get away with that? Many people had heard him argue vociferously in favour of the theory that the Earth had motion – not least His Holiness Urban VIII Barberini. And what punishment might he expect? Even though he probably still clung to the assurance of his age, fame and status, the thought of Bruno's fate and the grotesque, posthumous "punishment" of de Dominis only nine years previously, must have been in Galileo's mind.

Or perhaps his thoughts turned to Dante, his greatest compatriot, whose work he knew inside out. In the eighth circle of Hell the wanderer comes across cheats of various kinds, amongst whom is a certain Master Adamo, a forger who was executed in Florence in Dante's time. In one intense scene the sinner tells how he experiences eternity, mutilated and rooted to the spot, with a burning thirst and an unceasing longing for a single drop of water, a punishment for his "thirst" for wealth which led him to his offence.

Master Adamo was publicly burnt at the stake. It was no coincidence that the punishment for counterfeiting and heresy were the same. Both crimes represented attacks on the very foundations of society: the state's monopoly on fixing the worldly standard of value, and the Church's corresponding spiritual one.

When morning eventually arrived it was a weary old man that the guards came to fetch. They had with them a white gown which they put on him – the penitent's traditional garb.

Then Galileo was led out of the Inquisition's prison to a waiting cart. He was to make a public journey through the centre of Rome. The itinerary led him across the Tiber, through the narrow streets around Piazza Navona and ended close to the Pantheon, only a few paces from the very first place he had visited in Rome, 46 years earlier, the Jesuits' Collegio Romano.

But it was not at the Jesuits' that the old man in his white attire descended, but at their neighbours', the Dominicans. His journey ended in the small piazza in front of the austere, brownish-yellow brick facade of the church of Santa Maria sopra Minerva.

Galileo knew the church well, for it was closely associated with his home town. A statue of Christ by Michelangelo stood next to the choir, and the great Florentine painter Brother Angelico was buried there. But he was not led into the nave of the church with its wonderful sky-blue vaulting. He was taken through a side-door to the left, into the Dominicans' sober convent hall.

There, his judges awaited him: the heads of the Holy Office, the Council of Cardinals. But the Council was not complete. If Galileo had raised his eyes – his sight was no longer very good – to seek a glimpse of friendship or encouragement from Cardinal Francesco Barberini, it would have been in vain. The Pope's nephew was not there, and two other cardinals were missing as well.

Galileo was ordered on to his knees. Then the reasoning, judgement and sentence were read out.

The long sentences in flowing Latin that echoed under the hall's ceiling frescoes, were based on the most rigorous interpretation of the chain of events and the warning of 1616. True, the court did accept that Galileo might have forgotten the notorious words *teach* and *in any way whatsoever*, but thought that Bellarmine's written resumé, which Galileo had produced, far from served in his defence. Even if it did not contain the words *teach* or *in any way whatsoever*, it clearly stated that Copernicus' ideas were contrary to Holy Scripture.

In short, the publication of the *Dialogue* was "an open transgression of the said prohibition" (*aperte transgressio praedicti praecepti*). Galileo was thus clearly guilty, and unanimous judgement was passed "in the most holy name of Our Lord Jesus Christ" and also "that of the most glorious Holy Mother and everlasting Virgin Mary's". Galileo was found to be "vehemently suspected" of heresy and his *Dialogue* was banned.

From the opinion given, the sentence was actually quite lenient. Galileo's probable heresy would be forgiven provided he gave an immediate and public abjuration. To prevent any relapse and to emphasise the seriousness of the case he was further given a prison sentence "during our pleasure", and to read the seven penitential psalms once a week for three years.

Nothing was mentioned of the consequences of Galileo refusing to abjure. There were good reasons for supposing that such an eventuality would not arise. The ceremony continued as Galileo, still kneeling, was handed a document to read and sign. It began:

> "I, Galileo, son of the late Vincenzio Galilei, Florentine, aged seventy years, arraigned personally before this tribunal and kneeling before you Most Eminent and Reverend Lord Cardinals (...), having before my eyes and touching with my hands the Holy Gospels, swear that I have always believed, do believe, and by God's help will in the future believe all that is held, preached and taught by the Holy Catholic and Apostolic Church."

There followed a recapitulation of his offences, and then the abjuration itself: "I abjure, curse, detest the aforesaid errors and heresies and generally every other error, heresy and sect whatsoever contrary to the Holy Church." This was repeated in slightly different words twice more. The final part Galileo himself had to add to the document and sign:

> "I, Galileo Galilei, have abjured as above with my own hand."[100]

After this the ceremony was concluded, and Galileo was taken back to the Inquisition's rooms, which were now to be regarded as his prison.

Eternity

The most persistent myth regarding Galileo is that he rose from his kneeling position in the Dominican's hall and muttered obstinately: "*Eppur si muove*" – "it moves all the same."

The kernel of the bizarre ceremony that Galileo had just been subjected to, was the abjuration. The word "heresy" (*haeresia*) was in fact used by the Inquisition to mean two slightly different things: one was the pure denial of doctrinal truths, as when Lutherans regarded the Eucharist as a "symbolic" meal in which Jesus' body and blood were not literally present; the other was the transgression of Church commands or ordinances.

The teachings of Copernicus are not directly referred to as heretical in the judgement (only "contrary to Holy Scripture"), so it was the *flaunting of the warning of 1616* – the "the clear transgression of the said prohibition" – that was Galileo's heretical act.

For sentencing purposes however, the difference was not material. Whatever kind of heresy was suspected, the only lifeline was to abjure. Anyone who refused to do this would, by definition, be confirming their heresy, and the only solution left was the stake. But in order to be given the opportunity to save oneself by rescinding, the court had to be satisfied that the defendant, with his entire body and soul, wished to make good his errors. Torture was often used for this purpose, to get at the real truth of motives and attitudes. The abjuration then provided legally binding "proof" that the repentance was genuine.

But this also gave the ceremony another legal function. If the sinner was *again* taken for heresy in the future, there would be no way back. He would then have broken the binding oath that the abjuration represented, and death by burning at the stake was inevitable. The Inquisition had stopped Galileo's mouth – and his pen – for ever.

It is quite certain that that deep down he still *believed* in the theories of Copernicus. But just as certain is the fact that from then on he refrained from the least expression of it.

The judgement against Galileo was not only relatively mild, but in the narrow legal sense, totally unimpeachable. He *had* overstepped the decrees of 1616, no matter how they were interpreted, because he *had* presented the teachings of Copernicus as overwhelmingly probable in his *Dialogue*. No matter which way he put the case himself, it required no more than common literacy to see that.

Even so, there is no doubt that Galileo came out of the process a deeply disappointed and broken man. Alone amongst those present in the convent hall, he knew that the judgement was as monumentally foolish as it was legally correct. It locked the Catholic Church into a hopeless intellectual position as the deaf and blind denier of an ever more obvious physical fact, a position that would turn into one of the most painful problems in the long history of the Church.

For Galileo personally it was probably even worse in that the very foundations of the case resembled moral treachery from a man he had counted as his friend, Pope Urban VIII Barberini. After all, in happier times Maffeo Barberini had penned a eulogy to the mathematician and signed it *come fratello* – "like a brother". His attitude now was anything but fraternal.

What Galileo did *not* see was that he himself, with his self-assertiveness, impatience and provocative style, had made many enemies and thoroughly contributed to souring the atmosphere in Rome.

Perhaps another of the actors involved saw that the judgement was intellectually and morally bankrupt too, but he was not present in Santa Maria sopra Minerva. The Pope's own nephew, Cardinal Francesco Barberini, did not put his name to the judgement. This may have been accidental. The duty of attending plenary meetings was not taken all that seriously by cardinals, and two others were also absent. But Francesco had been a member of the Lyncean Academy, he understood the arguments in favour of the Copernican theories and knew they could not be magicked away by references to Holy Scripture and tradition. Furthermore, although he undoubtedly had a great deal to thank his uncle for, he had also witnessed at first hand the disconcerting alteration which had turned an open, intellectually inquisitive Maffeo Barberini into the suspicious, pompous and self-important Urban VIII.

Of all the people involved with the case, it was only Francesco who had known *both* Galileo and Urban VIII for many years.

What Urban's innermost thoughts were, nobody knows. But possibly he was more concerned with another, imminent occasion. Six days after the judgement was pronounced he consecrated Bernini's bronze baldachin in St. Peter's, a huge construction that was half sculpture and half architecture. Here, only the pope was – and is – allowed to conduct mass.

On the marble foundation of this definitive centre of the Catholic world were carved the Barberini bees.

The judgement in Santa Maria sopra Minerva was not just aimed at Galileo personally, it was also to put an end to the spread of Copernican ideas as a whole. To this end it was immediately copied and sent to the Inquisition's other offices around Italy and in the rest of Europe. Accompanying it was an instruction to local inquisitors to make the judgement commonly known, especially amongst mathematicians and philosophers. Soon acknowledgements began to arrive from all corners that the order had been followed.

The Inquisitor at Padua for example assured them that, not only had he made the judgement and revocation known to the professors of philosophy and mathematics at the university, but he had also included "other public lecturers", the priesthood, various scholars, "our writers" – and had a copy displayed in every booksellers.

On the other hand, he had not had much luck regarding the second part of the judgement: the banning of the *Dialogue*. The Inquisitor had only had one copy handed in, from a philosopher who clearly was too frightened to keep it any longer but, despite using his "very best efforts", he had not succeeded in getting hold of others. This was hardly strange: the book had immediately become a much sought after black market item, which was changing hands for twelve times its original price.

Galileo only served one night on the Inquisition's premises. The next day he was told that, for the moment, his incarceration could be transferred to the Embassy in the Villa Medici. It is likely that this was the work of Francesco Barberini.

But the Villa Medici was not intended to house prisoners indefinitely, nor did the Ambassador want to take responsibility for the shaken and despondent Galileo. An official request to the Pope that Galileo be allowed to return to Florence to serve his sentence there was, however, immediately turned down.

The solution was found in an unexpected quarter. The Archbishop of Siena, Asciano Piccolomini, belonged to a Tuscan family of considerable standing, which had produced both scholars and leading churchmen – the most famous of which was the Renaissance Pope, Pius II. If the august Archbishop's respect for the upstart Urban VIII was less than enthusiastic, his admiration for the Tuscan Galileo was all the more genuine. Piccolomini had read the *Dialogue* and had realised that Copernicus was probably right, and that the book would get its author into very hot water.

Now he offered to assume responsibility for this celebrated prisoner and hold him in house arrest at his palace. It was a suggestion nobody could object to. It got the problem of Galileo away from Rome and closer to his home territory, all the while reassuringly keeping him under clerical supervision. The Archbishop immediately despatched his own carriage, and on 6 July Galileo left the Villa Medici. For the sixth and last time in his life he bade farewell to Rome.

Archbishop Asciano of Siena was a wise as well as a learned man, and he understood human nature. He welcomed Galileo and reported back to Rome the very next day in an unusually reserved letter:

> "... yesterday Signore Galileo Galilei arrived at my house, to serve what has been ordered of him by the Holy Congregation, whose commands will be strictly obeyed by me, in this as in all other things. I am required to answer your Eminences in this way, and I humbly comply."[101]

After this very nominal obeisance on paper, the Archbishop forgot about all humility towards the Holy Office, and energetically put Galileo to the only thing that could set the broken down and sleepless old man on his feet once more: work.

The Archbishop's palace lay cheek by jowl with Siena's monumental cathedral, a piece of architectural artwork in light and dark green marble, that had once been intended to demonstrate to the main rival, Florence, the extent of the power and wealth the Sienese had at their disposal. But Piccolomini was now bishop of a tranquil provincial town, where only the great Palio horse race in the city square harked back to days of former glory and festivity.

In these peaceful surroundings the Archbishop exhorted his guest and "prisoner" to think about mechanics. Any sort of work on cosmological or astronomical problems was obviously completely out of the question. Galileo had to get on a new track or, more accurately, to return to the questions that had exercised him for fifty years: motion, fall, speed, acceleration – everything he had worked on so intensively, but had as yet written nothing about.

The judgement had not only limited Galileo's physical everyday life, through exile and imprisonment; most serious of all for the proud Tuscan was that the ruling from the Holy Office affected his honour and standing. To live in retreat as a "private person" had no meaning for a man of Galileo's background and ambition – his identity was linked to the social and public position he had attained. He wrote to his daughter Maria Celeste at the convent, saying that he felt as if he had been "struck from the rolls of the living".

There were just two things that might to some extent rehabilitate him in this position. The first was being allowed to return to Florence where he could have some contact with the court. His friends immediately began to work for this. But the more important was to write another book that would astonish the learned circles of Europe.

The Archbishop encouraged him all he could. Though the house arrest was formally maintained, his ecclesiastical gaoler ensured that interesting people were invited to his palace, so that Galileo might have discussions with them. Nor was Asciano Piccolomini particularly concerned to hide his true thoughts on the wisdom of the Holy Office's decisions. He expressed himself so openly and let Galileo have such a free rein that representatives of the more junior clerics of the diocese were soon sending an anonymous letter of complaint to Rome. In it, Galileo was accused of spreading "un-Catholic ideas" in the town, and of saying he could prove his philosophical hypotheses with "invincible mathematical reasonings". The Archbishop was personally accused of claiming that his prisoner was "the greatest man in the world", and that all progressive thinkers agreed with him.

In this atmosphere, some of Galileo's irrepressible optimism returned. He was still badly affected by rheumatism during the autumn, but more or less got over his insomnia and some uncontrollable spasms in his limbs which had begun after his humiliation in Rome. And so he decided to make the effort: Sagredo, Salviati and Simplicio were to meet again.

A Death and Two New Sciences

The Holy Office was clearly accustomed to anonymous written complaints. At any rate, it took no obvious notice of the accusations against Galileo and his most venerable host, unless the decision to allow Galileo to return to Florence was motivated by a feeling that it might be advantageous to get him away from an influential aid as powerful as an archbishop. If those in the most elevated circles of the Church reckoned that Grand Duke Ferdinando II was easier to control than one of their own prelates, they were completely right.

Formally, it was on grounds of health that Galileo's punishment was commuted to house arrest in his own villa in Pian' di Gulliari near Arcetri. But the terms were tough – he had to live alone, was not allowed guests except by permission of the local Inquisitor, nor, quite obviously, was he allowed to get involved in teaching or the discussion of cosmological subjects. By December 1633, Galileo was home again.

In practice the conditions of his house arrest were not so strict as to prevent him visiting his daughters in the nearby convent of San Marco. But a new worry awaited him there. His wise and practical daughter, Sister Maria Celeste, had informed and comforted him with her letters during the whole of his sojourn in Rome and Siena. She had diverted his thoughts to concrete, everyday problems: "... the reason the wine spoils is that you have never had [the casks] taken to pieces in order to expose the wood to the heat of the sun."[102] Maria Celeste had even promised to carry out the part of his punishment that consisted of the weekly repetition of penitential psalms, and was already busy with it.

But she was not well. The worries about her father's fate had affected her, she had stomach pains and felt weak and ill. In the spring of 1634, a month after her father's seventieth birthday, Maria Celeste got dysentry. Over the next week she rapidly weakened, and died peacefully in the convent on 2 April.

The sorrow of it almost broke Galileo. He put aside his work on the new book. Archbishop Piccolomini sent his condolences: "I have long known that she was the greatest blessing you had in this world," he wrote, and added comfortingly that she was now in a different and better one. Galileo described his own condition: "... my pulse is irregular because of disturbances of the heart, [I suffer from] deep melancholy and complete lack of appetite." He also described how, in his loneliness, he heard his daughter's voice calling to him.

A well known remedy for dejection and melancholy was wine, especially good wine. Luckily his friends realised this and sent him presents. He thanked one for sending him samples of two wines from the "wooded slopes that Bacchus loved" (it is not clear which district is being referred to): "They are different in taste, but of equal goodness and quality, and they ease my throat so much that I try to enjoy them alone, without sharing them with others." "A joyful mind," he went on, "is what best preserves life and health."[103]

The following year he received a magnificent gift from Grand Duke Ferdinando: more than a hundred bottles of wine from many different regions, and he mentions gifts of wine from "the Cardinal" (this may have been Francesco Barberini), the Grand Duke's younger brother and the Duke of Ghisa. Characteristically enough he had a special fondness for *siracusano*, the wine from the area around Syracuse in Sicily. Not only was this southern wine full-bodied and strong, but Galileo assumed it was the same wine which "my teacher Archimedes"[104] once enjoyed – the great philosopher and practical physicist had indeed lived in the Greek colony on Sicily.

During the spring and summer Galileo managed once more to find the strength to continue his book. His working conditions were not all that inspiring. No limitations were placed on his correspondence, only on his visitors, but the problem with his eyes which had been troubling him for some years gradually worsened. He was now amazingly independent of source literature to complete his new work. (Incredibly enough, Galileo's library consisted of only about forty books at the time of his death. His wine cellar was much better supplied!) It built to a very large extent on his own work of many years, but he had to be able to read his own notes. In addition, he felt his age, his ailments and the other pressures, and had to admit that, here and there, it was hard for him to follow the subtle reasonings he had sketched out in his younger days.

But the news that reached him from outside, also brought encouragement. The Holy Office had no authority in the France of Cardinal Richelieu. A copy of the *Dialogue* had fallen into the hands of an Austrian admirer of Galileo's, who translated it into Latin, the lingua franca of the learned. The translator ensured that the book was printed in Strasbourg – with the help of a Dutch publisher, the famous Louis Elzevier at Leiden. This was in 1635, and the following year Elzevier also published the *Letter to Christina*, in its Italian original with a Latin translation.

Elzevier had nothing to fear from printing this theologically controversial work. The Dutch had thrown off Spain and Catholicism and shone out as an oasis of liberalism in Europe, though admittedly some intolerance existed amongst extreme Calvinists.

As Galileo had predicted, the Copernican system had become thoroughly accepted in northern Europe, thanks in no little measure to his *Dialogue*. But the Inquisition's judgement was not without ramifications. It caused the pro-Galilean René Descartes – a devout Catholic who had been part of the Emperor's army early on in the Thirty Years War – to lay aside his finished work on the new world view, even though he lived in the Netherlands and was not in danger from any direct action.

Most important, however, was the enthusiasm the *Dialogue* created. The Dutch mathematician, Martinius Hortensius, had obtained a copy as early as the summer of 1634 and became an eager Copernican. In an inaugural lecture that year Hortensius went into the status of mathematics as a science and called it "a queen, reigning over man's spirit and actions".[105] Nor was it solely as an interpreter of Copernicus that Galileo influenced the scholars of Europe. His controversial hypothesis that "philosophy is written in this grand book (...) in the language of mathematics" also began to make headway to the great embarrassment of academic Aristotelians.

Galileo managed to get enough of his new book ready to begin to worry about the next problem: where to get it published. The judgement did not expressly say that he must refrain from publishing *anything* ever again; it only concerned his relationship with Copernicus. Pope Urban VIII was not finished with the matter, however. When, on Galileo's behalf, the Tuscan Ambassador asked if the old and infirm prisoner might have a dispensation from house arrest in order to visit a doctor in Florence, the retort was that unless such applications ceased, Galileo would be fetched back to Rome and put in the Inquisition's gaol there! And as for his books, the Pope decreed that *no* work by Galileo might be printed, not even reprints of books that had come out years ago.

At first it looked as if the Republic of Venice might be his salvation.

Courageous Paolo Sarpi had been followed by a worthy heir, Father Micanzio. He had written Sarpi's biography and, on his death, had assumed the position of theological adviser to the Venetian Senate, a position that entailed many confrontations with Rome as Venice still was not especially keen to bend the knee to the dictates of the Church in matters large and small.

The fearless Micanzio had known Galileo since his Padua days and was an undisguised supporter. During the case he wrote in a letter:

> "May that not disturb Your Lordship nor distract you from going ahead. The blow has been made: you have produced one of the most singular works that have been published by philosophical genius. To forbid its circulation will not diminish the glory of the author: it will be read despite the evil jealousy, and Your Lordship will see that it will be translated into other languages."[106]

But when it came to doing something for Galileo, Micanzio greatly overestimated Venetian, republican independence. He raised the matter with the local Inquisitor only to learn that The Lord's Prayer would probably have been denied an *imprimatur* if Galileo had been the one wanting to publish it!

With the aid of the Grand Duke investigations were carried out into the possibility of publication in the German language area. But here there was a detail that caused Galileo to give up the idea: Father Christopher Scheiner had returned to Germany. Jesuit influence was indeed large, but not all pervasive, so it is not certain what kind of fuss Scheiner could have started. However, it was obvious that Galileo had been so thoroughly frightened that he would not take any chances.

Much points to the fact that Galileo never *completely* took in the tremendous change in Urban VIII Barberini's attitude. As a consequence, he laid much of the blame for what happened at the door of the Jesuits. Several of his friends shared this opinion. In a letter, Galileo repeated something purported to have been said by the Jesuit mathematician Grienberger:

> "If Galileo had known how to keep the affection of the Fathers of this College, he would live gloriously in this world and none of his bad times would have come to pass and he would have been able to write as he wished about everything, even, I say, about the motion of the earth."[107]

However there is some doubt that this is a correct quotation from the otherwise cautious and discreet Grienberger.

While his friends were working on publication possibilities within Europe, Galileo got a surprising dispensation to journey more than thirty miles from his house. The probable reason for the Holy Office's tractability on this special occasion was the reason for the journey: the French Ambassador to the Papal States had expressed a desire to meet the ageing mathematician.

Such a request was difficult to oppose on purely diplomatic grounds especially now that the war in the north was entering a new phase which made Urban VIII's balancing feats even harder. Spain had beaten the Swedes and the other Protestant troops in southern Germany, and thus re-established a definite Catholic dominion. But this caused France to enter the war directly against Spain in 1635. From being a religious war, or at least a conflict with heavy denominational overtones, it had turned into a power struggle between the two leading Catholic states.

Ambassador François de Noailles had studied under Galileo in Padua, and was shocked at the treatment his old professor had received. Now he was on his way back to Paris for consultations about the turbulent situation, and would be passing through the little town of Poggibonsi, south of Florence. Galileo was given permission to meet him there.

The talk with de Noailles was a great encouragement to the isolated Galileo, and both men naturally talked about the publication prospects for his new book – the translation of the *Dialogue* had, of course, been printed in French territory. But it is less certain whether Galileo had brought a copy of his new manuscript with him as a gift for the Ambassador – as he later claimed. But the meeting furnished him with an admirable explanation for how the manuscript got out of the country – as diplomatic baggage!

The solution to the publishing conundrum lay in Holland. Louis Elzevier from the publishers in Leiden visited Italy and, with or without permission, also met Galileo, at his house. At the time the manuscript was not complete. Elzevier took some of it with him, and was to have the rest forwarded via Micanzio in Venice.

To Elzevier's mild discomfiture Galileo never seemed to be finished with it. The reason was simple. The old man with his failing eyesight realised that this would be his last book, and wanted to include all his thoughts and ideas – including the new ones which, even now in his dotage, never ceased to crowd into his mind.

This book, too, was written in dialogue form. Gradually, Galileo got four "days" ready, and had definite plans for a fifth; while simultaneously sending the publisher an "appendix", which had nothing to do with the rest of it.

Understandably enough, Elzevier got rather impatient with this method of working. Finally – in 1637 – the firm announced that they would print "four days" and the appendix, and requested a preface and dedication.

The situation was rather complex. Elzevier's firm was safely outside the reach of the Inquisition, but Galileo was not. So he came up with an elegant, if not entirely truthful solution to the problem. He could clearly not "present" the book to any Italian lay or ecclesiastical potentate by means of a dedication. And so he selected Ambassador de Noailles, well aware of his powerful position in France, where such a dedication really would be regarded as an honour. At the same time it would make it harder for the Church to interfere. It was inadvisable in the prevailing delicate state of foreign policy to do anything that might offend a prominent representative of France; especially as an attack upon Galileo would inevitably be seen as papal support for Spanish conservatism.

But the dedication also provided Galileo with a chance to deny all responsibility for the book's printing. As he himself described it, de Noailles had taken away a private, handwritten copy, and then suddenly "I was notified by the Elzevirs that they had these works of mine in press and that I ought to decide upon a dedication at once."[108]

As it turned out the Church tacitly accepted this fiction. Galileo's last book finally came out in Holland in 1638. It was in Italian and went under the title *Discorsi e dimonstrazioni matematiche intorno à due nuove scienze* – "Discourses and Mathematical Demonstrations about Two New Sciences". When a few copies gradually arrived in Rome and the rest of Italy, they were sold and read without Church interference.

The title, in fact, was Elzevier's and not Galileo's, something he regretted considerably, but nobody knows his own suggested title. In one way it made very little difference. By the time he held the book in his hands, he was no longer able to read anything at all.

The Meeting with Infinity

Galileo's last book is usually referred to by the simplified title *Two New Sciences*. This time the three friends, Salviati, Sagredo and Simplicio meet after visiting the Arsenal, Venice's famous shipyard. Salviati is impressed with all the *practical* knowledge accumulated there:

> "The constant activity which you Venetians display in your famous arsenal suggests to the studious mind a large field for investigation, especially that part of the work which involves mechanics; for in this department all types of instruments and machines are constantly being constructed by many artisans, among whom there must be some who, partly by inherited experience and partly by their own observations, have become highly expert and clever in explanation."[109]

The first of Galileo's "two new sciences" is an attempt at a technical treatment of the characteristics of matter, with special emphasis on fracture and deformation. The book opens with a discussion about the way a large ship is more prone to break up due to its own weight, than a small one built to the same proportions.

Salviati quite correctly states that one can raise a small obelisk without difficulty, whereas a large one of the same proportions is likely to fracture under its own weight.[110] This leads on to far more fundamental questions: what in fact holds matter together? How is it built up?

It is remarkable how much Simplicio's role has changed since his appearance in the *Dialogue*. He is no longer the naive and slightly unsophisticated Aristotelian who invites the sarcastic comments of the others. His function now is to act as an intermediary between Aristotelian physics and the mathematically orientated physics of Galileo. Whenever he introduces Aristotle's observations, they are received with deep respect by his conversational partners. Sagredo even quotes one "infallible maxim of the Philosopher".[111] This is a world away from the caustic criticism that Simplicio's hidebound intellectual conservatism unleashes in the *Dialogue*, where Aristotle's influence is regarded as the greatest bar to scientific progress. But in terms of literature, this sea-change causes the tension between the characters to slacken, making *Two New Sciences* a much less engaging read.

Salviati's speculative account of the construction of material is based on the traditional understanding (shared by Aristotle) that "nature abhors a vacuum".[112] He assumes that all material is composed of "atoms" – the smallest, indivisible entities of the material – which are held together by minute vacuums – *vacua* – which exert what might be called "negative pressure". This pressure keeps the material firm and intact, but there must be a great many such vacua in the toughest and least breakable materials. Galileo is actually calculating the weight of atmospheric pressure here, without realising it.

Through quite complex geometrical reasonings the disputants arrive at a point where they find themselves almost forced to their knees – before infinity. Simplicio jumps in and protests at the idea that a finite line has an infinite number of points along it – for, as he says, a long line must contain more points than a short one, but it is meaningless to say that one infinite number is higher than another.

Salviati demonstrates, in a most elegant way, that the concepts of "larger" and "smaller" cannot be applied to the infinite.[113] He takes numbers as his example. The amount of ordinary numbers is obviously infinite. But every number has a square ($2^2 = 4, 3^2 = 9, 4^2 = 16$ etc.). Thus the number of squares is *also* infinite – even though the sequence is clearly "smaller" because it does not contain the numbers in between the squares.

Galileo – through Salviati – does not stop here, even though he admits that our limited human intellect may not be able to grasp the infinite. He also indicates that there may be something midway between the finite and the infinite, and that there are quantities which can be described using whichever numbers one wants. The number of points on a line is perhaps just one such "halfway house".

In this, Galileo was fairly close to a truth that was first fully revealed 250 years later: in 1874 Georg Cantor proved that there were *several* classes of infinity. In fact, the points on a line belong to the class that *cannot* be "be organised" into a progression of numbers. But these speculations show the intellectual force that still lived on in the old prisoner of Il Gioiello, and how he had retained his appetite for approaching the most fundamental problems with reasoning and strict logic.

The continuation of his reasoning is, however, strange and not so easy to understand. Salviati points out that the distance between the squares becomes bigger the greater the square roots become. Therefore this cannot be "the road to infinity", on the contrary it becomes more and more distant as the numbers increase. And so, the only really infinite number is 1![114] It contains every power ($1^2 = 1, 1^3 = 1$ etc.)

Galileo was a thorough-going rationalist and mysticism of any sort was foreign to him. But just here it is tempting to think that his mathematics has rubbed up against the boundary of metaphysics. Beyond it lies infinity, which can be summed up in the number one. And who is the One who contains Infinity? Who can it be but God himself.

The three of them press on. Salviati discusses the problem of the speed of light. Does light spread instantaneously, i.e. infinitely fast, or just extremely quickly? He even suggests an experiment to decide the question.[115] (The experiment was not precise enough because the speed of light is so great. But in a sense Galileo contributed when the speed of light was measured for the first time, by the Danish astronomer Ole Rømer in 1676. The measurement was done using the satellites of Jupiter.)

But this is the end of uncertainty and vagueness. The remainder of the first day is a veritable scientific triumph through the laws of motion, in which all of Galileo's experiments with free fall and pendulums are presented and summed up in exemplary fashion. Simplicio with his Aristotelian counter-arguments is amicably and respectfully put in his place. Especially masterly as a piece of scientific prose, is the long section where Salviati argues a proposition that seems quite improbable to Simplicio, namely that a wisp of wool and a lead ball will fall at exactly the same speed in a total vacuum.[116]

Salviati provides a careful account of air resistance. He also attempts to calculate the buoyancy of air, clearly based on experiments. The lack of accurate measuring instruments makes his estimate relatively imprecise. Salviati assumes that water is 400 times heavier than air, the correct figure being approximately 780 times. Otherwise his reasoning is so elegant and convincing that Simplicio announces that, if he were about to begin his studies afresh, he would start by reading mathematics!

The first day concludes with a section on pendulums, in which the vitally important law that states that the oscillation time of a pendulum is proportional to the square root of its length, is thrown in almost as an afterthought.[117] Much more space is dedicated to a fairly long exegesis about musical theory, which uses his experience of pendulums on swinging strings.

Galileo is here extending and rounding off the work of his father, with precise observations on the relationship of the strings' weight, length and taughtness – and the tones that result. It is all brought to a conclusion by Salviati explaining which intervals sound sweet to the ear and which jarring. The cause is purely mathematical:

> "The pulses delivered by the two tones, in the same interval of time, shall be commensurable in number, so as not to keep the ear-drum in perpetual torment, bending in two different directions in order to yield to the ever-discordant impulses."[118]

Put briefly: the number of vibrations (and thus the tone) must have a harmonic relationship, e.g. 2:3. Everything is a matter of proportion – even musical harmony.

At this point the three take a well deserved rest until the next day.

The second day offers a fairly brief and technical description of how one calculates the breaking strength of various bodies. Salviati returns to the starting point of the conversation, and demonstrates geometrically why large constructions are proportionately more vulnerable than small ones – and he explains that is why giants, many times bigger than ordinary people, cannot exist. If they did, their joints at least would have to be made of some other material![119] Yet again Simplicio is there with solid, sensible objections: whales, he says, are monstrously large.[120] And so Salviati gets the opportunity to elucidate on the effect of buoyancy; a recurring theme in Galileo's thinking throughout his life.

But it is the third day that is the most important in *Two New Sciences*. During it, the full panoply of the other new science is presented: the science of motion, kinematics. No longer is Galileo so concerned about maintaining the fiction of the three interlocutors. The chapter opens with a short dissertation in Latin – in his own name. The introduction is like a triumphal fanfare for the work of a lifetime:

> "My purpose is to set forth a very new science dealing with a very ancient subject. There is, in nature, perhaps nothing older than motion, concerning which the books written by philosophers are neither few nor small; nevertheless I have discovered by experiment some properties of it which are worth knowing and which have not hitherto been either observed or demonstrated."[121]

During the conversation that follows, Simplicio and Sagredo begin to discuss the *reasons* why objects move. Salviati interrupts them politely but firmly by saying that there are many causes: "the attraction of the centre" (the force of gravity), the influence of the medium they are moving in, a force acting between the basic elements within the object.[122] But, he says, Galileo's method is to investigate and show *how* motion occurs, not why.

Salviati is thus encapsulating something of what is most crucial in Galileo. His mathematical mode of thought provides a description of *what actually happens*, whereas the traditional Aristotelian logic was always concerned with speculations about cause and effect, without grounding itself in a sufficiently stringent description of reality. Even Simplicio eventually begins to understand a little of this when he extols the exactness of mathematics.

This is the only place in *Two New Sciences* where the suspicious reader might capture an echo of the debate surrounding the Copernican system. Salviati speaks about what can happen to someone who puts forward irrefutable proof that old, deep-rooted notions are erroneous:

> "[There is] a strong desire to maintain old errors, rather than accept newly discovered truths. This desire at times induces them to unite against these truths, although at heart believing in them, merely for the purpose of lowering the esteem in which certain others are held by the unthinking crowd."[123]

Wisely, Galileo does not continue along these lines. But he clearly believes that this description would fit a Grassi, a Scheiner – and maybe even an Urban VIII – well.

The fourth day also concerns motion – but this time "forced", not "natural" motion (like free fall). The starting point is the sadly practical application such investigations have in ballistics. Musket and cannon balls fly in this chapter. By means of elegant conic section geometry, Galileo (Salviati is now again merely a commentator) proves that, if the ball is fired horizontally, its trajectory is parabolic – provided one accepts his assertion that the ball's curved line of movement can be analysed as consisting of two entirely *independent* motions. One is the even movement on the horizontal plane imparted by the power from the weapon, the other is the free fall which affects all bodies.

This insight is possibly as important as the law of fall. It forms the basis of all practical descriptions of actual motion.

Salviati promises that the three of them will meet again to talk about impact, that is, brief contact between bodies – on the work's fifth day. But it was never written.

After that *Two New Sciences*, and Galileo's scientific output, end with the appendix. This comprises a fifty-year old paper about the centre of gravity in bodies. The old prisoner in Il Gioiello delves into the thoughts of the

23-year old who went to Rome and disputed with the Jesuits. And so his life's work is brought to a close.

"That Universe . . . Is Not Any Greater Than the Space I Occupy"

> ". . . alas, my lord, your dear friend and servant Galileo has for the past month become irreparably blind. Now imagine, Your Lordship, how afflicted I am as I think about that sky, that world and that universe which I with my marvellous observations and clear demonstrations had opened up hundreds and thousands of times more than had been commonly seen by the sages of all bygone centuries; now for me it is diminished and limited so that it is not any greater than the space I occupy."[124]

Galileo dictated these words on 2 January 1638.

Two New Sciences was a publishing success. The book aroused particular interest in Germany and France, and a French translation was ready within a year. But fifty copies also found their way to Rome, without anyone trying to prevent it – in fact, Cardinal Francesco Barberini bought the book himself. As it had been printed outside the Inquisition's jurisdiction and, as it patently did not contain a trace of Copernicanism, the book was left alone and quickly sold out.

A copy eventually found its way to Arcetri and the Villa Il Gioiello. But by the time Galileo had the book in his hands, he was completely blind.

Galileo worked on with intensity, both while his sight slowly dimmed and after he had become enveloped in total darkness. The prohibitions that circumscribed his freedom of action were never officially lifted, but gradually things became somewhat easier. Two young pupils moved in and did his letter writing and read aloud to him: first, the barely sixteen-year old Vincenzio Viviani, then in Galileo's final months the older and later more renowned Evangelista Torricelli, who took up the torch of the master's ideas concerning atmospheric pressure, and constructed the first barometer.

Before he lost his sight, Galileo managed to make one last important astronomical observation using his beloved telescope. He had studied the Moon for more than 25 years. No other person knew every detail of its surface as he did. Now he realised that occasionally it was possible to see small areas that usually did not form part of the visible area. He was able to determine that this heavenly body displayed a minute "rocking" movement when seen from the Earth. He called the phenomenon the *libration* of the Moon.

The letter he wrote to Father Micanzio in Venice about this libration contains a couple of noteworthy sentences. Galileo wonders if the phenomenon can have an effect on the tides! And as if that were not sufficient, he rounds off with:

> "... [the tides] which by the common consent of all, the moon is the referee and superintendent."[125]

In this little clause lies the undermining of the entire fourth day of the *Dialogue*. In the quietness of his isolation, Galileo had begun to doubt his great idea, his epoch-making and definitive proof of the Copernican system. He had – quite contrary to his normal instinct – begun to move towards "the common consent of all" [*comune consenso di tutti*].

It is hard to conceive that this happened for religious reasons, or out of respect for the Inquisition's judgement. Maybe Kepler's argument had finally made an impression on him, or perhaps he had simply gone through his own reasonings once more and seen the weakness of them. If so, this was one of his greatest intellectual achievements: it is one thing to see through the failing arguments of others, quite another to examine a line of reasoning that has formed a central plank in one's own view of the world, carefully and critically.

Another old project close to his heart, still took up a lot of Galileo's time. This was fixing longitude by exploiting the satellites of Jupiter – the characteristic combination of cutting-edge science and sober, practical application.

Galileo's great admirer in Amsterdam, the mathematician Hortensius, was given the task of working out the possibility of securing the rights of this for the Netherlands. He was undaunted by the practical difficulties the observations posed. Hortensius contacted Galileo, and had planned a journey to Italy. But suddenly this talented man died at the age of just 34, in 1639. This put paid to the maritime use of the Medicean stars for good.

The blind old man at Il Gioiello was one of the most famous men in Europe, and despite the embargo on visits, colleagues and admirers came in secret to pay their respects. One of them was John Milton. That great and thoroughly learned English poet was deeply interested in astronomy. In his principal work *Paradise Lost* the struggle between God, Satan and the angels is set in a carefully constructed universe, although largely built on Ptolemaic principles, mainly out of poetic considerations. Milton (who would also live to be blind and isolated in his old age) used the experiences of his meeting with Galileo in a domestic political context. He was one of Cromwell's supporters and emphasised the relative freedom of thought that existed in England, as opposed to the Catholic intolerance that had affected Galileo so deeply.

However, Galileo also received discreet assistance and support from the Church as well.[126]

The so-called Piarist movement was a peaceful order of kindly, learned brothers. The order was well represented in Tuscany and enjoyed the especial goodwill of Grand Duke Ferdinando. The Piarists ran what was called *scuole pie* – "pious schools" but, in contrast to the powerful and intellectually aristocratic Jesuits, they worked quietly at the grass roots, giving elementary education to poor children in reading, writing and arithmetic.

In certain places – including Florence – the Piarists had also begun to give higher education, in a very modest and low key manner so as not to upset the Jesuits. The Grand Duke was so pleased with this that he allowed one prominent Piarist to teach two of his younger brothers. He also looked favourably on the fact that, in practice, Father Clemente Settimi took on the role of Galileo's secretary. It was to Settimi that the indefatigable Galileo at an age of 76 dictated a letter containing his thoughts on the *cycloid*, the curve described by a fixed point on a circle that rolls along a line.

Father Clemente also functioned as Galileo's nurse. It was against the rules of the order to spend the night outside the cloister, but the monk got special permission from Rome, so that he could stay with Galileo when necessary.

It was not merely pious humanity that drove the Piarists in Florence to help Galileo. Settimi and other brothers with a mathematical education were in reality convinced Copernicans. But the Inquisition had merely relaxed its grip a little in regard to Galileo; it certainly did not slumber. And so Clemente Settimi's efforts for the old man came to an abrupt end when a member of the order reported his mathematical colleagues to the Holy Office:

> "All the above maintain that there is no truer or surer science than that which Galileo teaches with the help of mathematics; they term it new philosophy and the true way to philosophise, and they have many times said (...) that it is the true way to learn to get to know God (...)"[127]

Neither the Pope nor the Holy Office could tolerate such ideas spreading within the Church, no matter how devout the Piarist Order was in its day-to-day running. The order was dissolved in 1646.

The person who was closest to Galileo in his last years, apart from his young pupil Viviani, was his son. The clashes that had soured their relationship in Vincenzio's younger days were gone. His son had become a responsible family man, and it was to him that Galileo confided his very last project.

It was quite unnecessary to use the satellites of Jupiter to establish longitude. The only advantage they had was that their eclipses were many and predictable – they were simply to be used to fix the time accurately. The whole thing could be done much more simply if one could just construct a temporal timepiece that was absolutely dependable.

Galileo may have had the idea earlier but in 1641, at the age of 77, he tried to breathe life into it. Sixty years earlier he had discovered that small pendulum movements were a sort of measurement of time, but he had used this for nothing except his curious *pulsilogium*. Now he knew a lot more about the characteristics of the pendulum, and realised that theoretically it could be used as the heart of a machine for measuring time – a perfect clock. But he could not sketch out the principle himself, far less construct a working model.

So he summoned his son and explained the idea, an idea which might secure the family's future prosperity if it could be manufactured. But Vincenzio was no enterprising innovator like his father. He let the project lie, and so it was the Dutchman, Christiaan Huygens, who eventually made the first working pendulum clock in 1656, in perfect keeping with the shift of science and technology to northern Europe.

In the autumn of 1641 the 33-year-old Evangelista Torricelli came to Il Gioiello. It was he who was to take down Galileo's last thoughts.

The Grand Duke's old mathematician turned back to Euclid at the last, a companion throughout his long life and the very foundation of his attempts to "read" nature – "this grand book (...) which stands continuously open to our gaze".

In the fifth book of Euclid's *Elements* the general rules of proportionality are defined, both those of arithmetical quantities (numbers) and geometrical ones (areas, bodies). Sick and bed-ridden, unable to make a note or open a book, Galileo dictated his new interpretation of certain passages that had always given students of Euclid problems. He still remembered his old friends Sagredo and Salviati, who had been dead for more than twenty years, because he dictated in dialogue form.

But his strength was not up to it. On the evening of Wednesday 8 January 1642, barely a month before his 78th birthday, Galileo died in his bed. With him were Vincenzio, Torricelli and Viviani, who was to write the first biography of Galileo.

In it he describes the death scene:

"With philosophical and Christian constancy, he [Galileo] rendered his soul to his Creator, sending it forth, as far as we can believe, to enjoy and admire more closely those eternal and immutable marvels, which that soul, by means of weak devices with such eagerness and impatience, had sought to bring near to the eyes of us mortals."[128]

Epilogue

Pope Urban VIII also displayed constancy when he heard the news of Galileo's death, but it was not of the philosophical or the Christian sort. The news reached Rome in a letter to Cardinal Francesco Barberini from the Nuncio in Florence (who had obviously been given the wrong date):

> "Galileo [Il Galileo] died on Thursday the 9th, on the following day his body was privately placed in Santa Croce [Church of the Holy Cross in Florence]. The word is around that the Grand Duke wishes to provide a sumptuous tomb for him comparable to and facing that of Michelangelo Buonarroti and he is of a mind to give the modelling of the tomb to the Academy of the Crusca. Out of my respect for you I thought that Your Eminence should know this."[129]

The Nuncio's respect for Francesco's uncle, who was the real intended recipient of the rumours, was even higher. And His Holiness' view of Galileo was unchanged, as the Tuscan Ambassador was to find out in an audience. This was his report home to Florence regarding his conversation with Urban VIII, a true study in the art of diplomacy:

> "... he told me that he wanted me to share with him in confidence a particular and only for the simple purpose of conversation and really not that I should be obliged to write anything about it; it was that His Holiness had heard that the Most Serene Master [the Grand Duke of Tuscany] may have had plans to have a tomb for him erected there in Santa Croce, and he asked me if I knew anything about it. In truth I have heard it talked about for many days now, nonetheless I answered that I did not know anything about it. The reply from His Holiness was that he had heard some news, but did not yet know whether it was true or false; at any rate he nonetheless wished to tell me that it would not be at all a good example to the world that His Highness would do this thing, since he [Galileo] had been here before the

> Holy Office because of a very false and very erroneous opinion, with which he had impressed many others around here, and had given such universal scandal with a doctrine that was condemned."[130]

Grand Duke Ferdinando was as usual keen to avoid unpleasantness, and Galileo's body was laid to rest in a modest side chapel, without an inscription.

To correct this and raise a worthy monument to his teacher became the life's work of Vincenzio Viviani. As he was obviously going to get nowhere with a physical marble tomb in the short term, he decided to preserve Galileo's memory in two other projects. One was an edition of Galileo's collected works (admittedly without the *Dialogue*). This he had ready by 1656. The second, and the more important, was his biography which was begun at about the same time, but not printed until after his death.

Being unable to position Galileo in the right place in the physical sense – right opposite Michelangelo – Viviani managed to do it in an intellectual, or rather a spiritual, manner. He simply moved Galileo's birthday forward three days, from 15 to 18 February 1564. It was on that day that Michelangelo died!

So Galileo was placed in the line of great Tuscans from Dante onwards. It was also popular at the time to compare him with Columbus. But in Florence a son of the city was naturally more appropriate than the Genoese Columbus, so Viviani drew the parallel with Amerigo Vespucci, the man who quite accidentally came to bestow his own name on America:

> "... the immortal fame of that other Florentine Amerigo, who not only discovered a piece of land, but innumerable worlds and new lights in the sky."[131]

But the other camp in the conflict also had their image of Galileo.

During the seventeenth century several Jesuits published historical reviews of the development of mathematics and related sciences. It was quite hard for them to avoid a man who was probably the greatest, and without doubt the most famous, scientist of the early part of their own century. They solved the problem of Galileo by distinguishing between his scientific achievements on the one hand, and his defence of Copernicus on the other. As a scientist he could be praised, more or less critically; as a Copernican he had inevitably to be condemned.

Urban VIII came up against many other problems during his long pontificate, and he did not always get his own way. His most painful setback came when he tried – on behalf of the Barberinis – to take the small Dukedom of Castro from their arch-rivals, the Farnese family. He did not balk at excommunicating the Duke as part of his power politics. But the other Italian states intervened, and Urban had to accept the previous status quo.

Pope Urban died in 1644. *His* monument was designed by Bernini and placed in St Peters as centrally as possible, right next to "the chair of St Peter". As was the custom, Urban himself had planned the monument. He wanted to emphasise his Florentine background, so Bernini was given an artistic pattern – Michelangelo's monuments to the Medicis in Florence. Urban VIII's grave was decorated with two allegorical statues, Charity and Justice.

Vincenzio Viviani went to his grave at the age of 81 in 1703, without having erected a monument. In fact – he went to *Galileo's* grave, because at his own request he was laid to rest in the same crypt as his old teacher. His will had a codicil that enjoined its beneficiaries to work for the good of the monument.

Galileo's life was closely linked to the Medici family, but their sad demise only enhanced his reputation. The last Grand Duke, Gian Gastone, inherited the throne because his elder brother Cosimo III died prematurely of syphilis. As for Gian Gastone, he slowly ate and drank himself to death. He had no heirs, and the great powers of Europe nominated the Habsburg Duke of Lorraine as his successor, without consulting either Gian Gastone or any other Tuscan. This in practice left Tuscany a mere appendage of the Austrian Empire by the 1730s.

Then a wave of nationalism swept the neglected and impoverished Grand Duchy. Sorrowfully, many compared the current dismal situation with the times when Florence and Tuscany were a cultural and economic focal point in Europe. It was a national uprising completely devoid of power, it was limited to symbols. To get Galileo's tomb sited directly opposite that of Michelangelo was a worthy symbolic gesture: then every visitor to Santa Croce would, immediately on entering, pass between these two figureheads of Tuscan art and science.

More than a century had passed since the trial. Pope Clement XII Corsini, was himself a Florentine, and had no qualms about a circumspect rehabilitation of Galileo to cast a bit of much needed glory over his native city. In the artistic sphere, too, things were definitely on the wane in Florence, but the sculptor Foggini who was given the commission, was a competent late Baroque master. Galileo is depicted in an heroic sky-gazing attitude, telescope in hand. Two allegorical female figures adorn his tomb as well; they are Astronomy and Geometry.

Galileo's epitaph was written in Latin, the only language formal enough for such an occasion, but it was perhaps a little paradoxical for a man who had consciously written his major works in his mother tongue. Galileo was described as "the restitutor of geometry, astronomy and philosophy, unparalleled in his age".

In 1737, on 12 March – the same date that Michelangelo had been interred in Santa Croce – Galileo's remains were transferred to a vault in the new monument.

At last Florence had restored the honour of its son. The fact that he was a Florentine was clear from the suffix to his name on the epitaph: "Patric. flor." – "notable citizen of Florence."

It took longer in the rest of Italy. In 1820 a professor at the University of Rome wanted to publish a textbook that explained the Copernican system without dealing with it as hypothetical – a view which by that time had clearly long been universal in professional circles. But a zealous clerical official refused to sanction the publication, citing the 1616 decree. This led to the most bizarre situation as the Holy Office had to step in and ensure the book was published, by threatening reprisals *against* the forces hindering the publication of an up-to-date textbook!

This meant that the formal grounds for a ban on Copernicus' *De Revolutionibus Orbium Coelestium* and Galileo's *Dialogue* were gone. When eventually a new edition of the *Index librorum prohibitorum* came out in 1835, both titles had quietly been deleted from the list.

Galileo became an important symbol for the forces that were working towards the unification of a fragmented Italy during the 19th century. This inspired Antonio Favaro to publish a scientific edition of his collected works in twenty weighty volumes, from 1890 to 1909. The importance with which Galileo was endowed can be seen from the series title of the volumes: *Edizione Nationale* – "The national edition".

During the course of the 19th century the Vatican's archives were opened to some extent to researchers who wanted to study Galileo. (The archives of the Holy Office have remained closed to this day, although certain documents have been made public after special application.) This led to a wave of ecclesiastical self-criticism, but it was a wave that gained momentum very slowly indeed. Even at the time of the great Second Vatican Council in the 1960s – a radical attempt to think through the relationship between the Church and the modern world – the Galileo affair was only alluded to in very vague terms.

It was Copernicus' own countryman, the Polish Pope John Paul II Wojtyla, who really got to grips with the Galileo problem in all its ramifications.

He has admitted the Church's errors on several occasions, for example during a speech he made at the University of Padua in 1992, where he was a guest at the 400th anniversary of Galileo's appointment as a professor there.

In the middle of the old city in Florence, only a few steps away from Cosimo I's Uffizi building, is the Palazzo Castellani, which nowadays contains the Science Museum – *Museo di Storia della Scienza*. Just as in the Uffizi Gallery, large parts of its collection came originally from the Medici's private collections.

Galileo is the museum's great attraction. The collections and the library can be regarded as a Mecca for modern research on, and interest in, Galileo. But amongst the collection of artefacts which belonged to, or can be associated with Galileo, is one rather peculiar object.

This is an egg-shaped glass container, decorated with a gilded metal band. The container stands on a tallish, cylindrical pedestal of marble and if you go close enough, you can see a lengthy inscription encircling the pedestal. The whole thing is about twenty inches high.

But what is clearly meant to be the main focus of attention is an elongated, slightly bent, greyish-white thing within the glass egg.

Galileo's lasting contribution to science is based first and foremost on his description of evenly accelerating motion (free fall), and on the early investigation of the heavens with technical aids. In both these fields he produced work that make him a key figure in the history of science. As important perhaps is his *approach* to natural science, with its emphasis on experiment, observation and mathematical processing, rather than tradition and abstract reasoning.

The paradox is that, although he is *remembered* more than perhaps any other scientist up through the ages, this is because of the battle over the Copernican system. In this revolutionary view of the universe he is, however, despite the breakthrough of his discovery of the satellites of Jupiter, little more than a footnote between Kepler and Newton in historical terms.

The dramatic court case with its plainly drawn battle lines turned him into a perfect symbolic figure. Together with his role as an experimenter – graphically brought to life in Viviani's story about him dropping balls from the leaning tower of Pisa – the case made Galileo seem like the father figure of modern science, a science that defies prejudice and stupidity in its pure search for knowledge. This is how he is described in a Norwegian sixth form college physics textbook from the 1960s:

> "His importance to science can hardly be overestimated, he must be considered one of its greatest men. He made the experiment the vital thing (...)"[132]

In Italy this has endowed Galileo with the status of a sort of secular saint, a symbol of intellectual freedom and rebellion against hidebound religious authorities. At least one Italian historian has been completely ostracised because he questioned the black-and-white attitude to the great man's martyrdom. And the feeling is not limited to Italy. In 1959 Arthur Koestler wrote *The Sleepwalkers*, in which Galileo's life and fate plays an absolutely central role as the very "watershed" where religion and science unhappily went their separate ways. He places a good deal of the failure on Galileo's difficult temperament. The book aroused violent and relatively unacademic reactions from the two leading Galilean researchers:

> "[The treatment of] Galileo is simply dishonest from beginning to end. (...) Koestler has threaded together every discredited charge, ancient and modern, that has been made against him [... and] added a few deliberate distortions of his own."[133]

Galileo's courageous and obstinate achievements in disseminating the new truths cannot be doubted. Nor does it diminish him in the least that his motives were mixed, as all human motives are, and that his intense stubbornness could turn into another human characteristic, self-righteousness. But the story of his reputation also shows that even people who distance themselves from religion on the grounds of rationalism, or at least reproach it for interfering in areas of life that are none of its business, also have need of saints and martyrs.

For such devotees Galileo is above any hint of criticism, he has become an icon, a character that is not to be sullied. We are forced to call such admiration worship. The strange object in the Science Museum in Florence emphasises this. For it is a worldly relic – an anti-relic, if you will, in a country whose innumerable churches are awash with sacred objects.

It is Galileo's right index finger, the finger that once clasped his pen and steadied his telescope when he turn it skywards.

Postscript

The literature on Galileo is enormous and fast-growing. It is also pretty diverse: Volker R. Remmert lists *eleven* different trends in modern Galilean research. Although my book is based on many sources, I did not want, nor was I able, to do justice to this plethora of interpretations. Instead, I have tried to draw a clear and complete picture of Galileo, his environment and his destiny. This, of course, has involved a number of choices along the way.

The following have had a strong influence on the background to my work:

Arthur Koestler's masterful, vivid and ever so slightly malicious portrait of Galileo in *The Sleepwalkers*. Half science, half novel, it needs a lot of amplification and correction, particularly in the light of all the new research that has been done in the years – more than forty of them – since Koestler wrote the book. That aside, it is the account that best whets the appetite for Galileo the *man*, and the role his personality played in his own destiny and that of his work.

Where Galileo's scientific contribution is concerned, I have largely followed Stillman Drake's elegant reconstructions, which place his fundamental discoveries on the theory of motion in the Paduan period. Drake is an irrepressible Galileo admirer and apologist but, as far as a layman can judge, his up-rating of Galileo as an experimental physicist is solidly supported by modern research.

In the past twenty years two books have appeared that provide wholly new interpretations of Galileo's career and destiny. Mario Biaglioli's *Galileo, Courtier* deals with the relationship that Galileo and the science of his age had with the complicated social structure surrounding the Church, the universities and the nobility – what we might call the patronage culture. Biagioli has had an influence on almost everything written on Galileo over the past few years. There are many traces of him in my work as well.

Pietro Redondi's *Galileo, Heretic* re-interprets the entire case, indeed even the course of events from early in the 1620s, in a sensationally novel way. He believes it was Galileo's atomism as formulated – almost in passing – in *The Assayer*, that was the real cause of Galileo's downfall, because that threatened to contravene the doctrine of transubstantiation. (Grassi's book from 1626, p. 160, would in that case take on an entirely different meaning.) Under these circumstances the court case of 1633 would appear to be some kind of manoeuvre to save Galileo from a more dangerous charge and certain conviction as a heretic. If Redondi is right, the history of Galileo's renunciation and his entire relationship with the Church, must be rewritten. However, Redondi has found little support amongst professional historians of science, and I have chosen not to base my work on his interpretation.

My account of Galileo's relationship with the Church relies on many sources, but most of all on Annibale Fantoli's *Galileo – For Copernicanism and for the Church*. This discerning and balanced exposition is part of the Vatican's new series of Galileo studies (*Studi Galileiani*), and must of course be read as such, but Fantoli's well documented and sober book provides a summary of what we now know took place on the stage and behind the scenes in the long, sad story of the Catholic Church's treatment of Galileo.

By far the most controversial point in the entire "Galileo case" is the memorandum from Cardinal Segizzi, the one containing the famous words *Nec quovis modo teneat, doceat aut defendat*. Until very recently it has been usual to assume that the document was a forgery, produced in 1632, perhaps directly at the behest of Urban VIII, to ensure that Galileo was found guilty. However, there seems little doubt that the document is genuine (see Fantoli, pp. 219–222). The details of what actually happened at the meeting with Bellarmine and Segizzi on 26 February 1616, which resulted in two different memoranda, is uncertain. My account follows Fantoli.

The use of the words "science" and "scientist" in my book may well be anachronistic. The terms should have been "natural philosophy" and "philosopher". But as we now regard much of Galileo's work as pioneering in that area of human cognitive experimentation *we* call "science", I considered that this would make the reading easier.

Galileo and his contemporaries themselves used the term "Italy". On the whole I have tried to avoid the word, however, and instead used "the Italian mainland", "the Italian states" etc., to avoid associations with the current Italian national state. The reason for this is that Italy's former division into lesser states and principalities plays a vital role in understanding the age, and the fate of Galileo.

The great ecclesiastical meeting which fired the starting gun for the Counter-Reformation was often called the Tridentine Council. However, *Trento* is the modern name for the town where the meeting was held, and I have therefore chosen to use "the Council of Trent" which accords with modern historical usage as far as I am aware.

In the main I have indulged myself in the use of the antiquated term "specific gravity" instead of "mass density", simply because I believe it to be better known.

The quotations from the Bible are from the King James version.

My thanks for help with this book go first and foremost to my editors, the tireless Hans Petter Bakketeig and Arne Sundland at Gyldendal Dokumentar, thereafter to everyone else with whom I have had conversations and discussions. Thanks also to the staff of Fredrikstad Library who have assisted me with many special book requests, to the helpful staff at the *Center för vetenskapshistoria* in Stockholm and to Daniela Pozzi of the *Istituto e Museo di Storia della Scienza* in Florence. Any misunderstandings or misinterpretations in this book, are not the fault of these kind helpers, but wholly and completely my own.

Appendix

Popes

Pius IV de' Medici	(1559–1565)
Pius V Ghislieri	(1566–1572)
Gregory XIII Boncompagni	(1572–1585)
Sixtus V Peretti	(1585–1590)
Urban VII Castagna	(1590)
Gregory XIV Sfondrati	(1590–1591)
Innocent IX Fachinetti	(1591)
Clement VIII Aldobrandini	(1592–1605)
Leo XI de' Medici	(1605)
Paul V Borghese	(1605–1621)
Gregory XV Ludovisi	(1621–1623)
Urban VIII Barberini	(1623–1644)
(...)	
Clement XII Corsini	(1730–1740)
(...)	
John Paul II Wojtyla	(1978–)

Popes mentioned in the text are italicised.

Grand Dukes of Tuscany

Cosimo I	(1537–1574)
Francesco	(1574–1587)
Ferdinando I	(1587–1609)
Cosimo II	(1609–1620)
Ferdinando II	(1620–1670)
Cosimo III	(1670–1723)
Gian Gastone	(1723–1737)

References

References comprising a Roman numeral followed by a page number are to *Le Opere di Galileo Galilei* (the Roman numeral indicates the volume). In quite a number of instances I have found it expedient to give two references to direct quotations: first to an English or German language source and then to *Le Opere*. In such cases SN refers to Albert von Helden's edition of *Sidereus Nuncius*. D stands for Stillman Drake's edition of the *Dialogue*. TNS is Crew/de Salvio's edition of *Discorsi e dimonstrazioni matematiche, intorno à due nuove scienze*.

A few quotations have been taken from von Gebler's critical edition and not compared with *Le Opere*.

All translations not otherwise credited are my own (and the English translator's). When translating from Italian to Norwegian (and a couple of times cautiously from Latin to Norwegian), I have sought support in English or German translations. English language translations have not been available to the translator in all cases, and he has therefore, occasionally, based his translation on my own.

1. Galileo (Rome 1623): *The Assayer*. In S. Drake and C.D. O'Malley: *The Controversy of the Comets of 1618*. University of Pennsylvania Press. Philadelphia 1960, p. 189; *Il Saggiatore*. Feltrinelli. Milano 1992, p. 48.
2. IX, 213–223.
3. Here quoted from Ginzberg, ix.
4. Ricci, 542–543.
5. Ricci, 545 ff.
6. Yates, 208.
7. X, 56–57.
8. Hamel (in von Maÿenn I), 136.

9. See Favaro's two pamphlets on the subject. *Of course this did not refer to* Gustav Adolf, *a claim that is sometimes made.*
10. Bassani/Bellini, 130.
11. XIX, 218.
12. XIX, 219.
13. XIX, 220.
14. X, 143–154.
15. II, 519.
16. Galileo Galilei: *Il Saggiatore*, 1623. Feltrinelli. Milano 1965, p. 38; Drake translation in *The Controversy of the Comets of 1618, op cit.*, pp. 183–4.
17. Bassani/Bellini, 233.
18. Wootton, 128.
19. SN 30–31; III, 56.
20. SN 93; X, 343.
21. Fantoli, (transl. Coyne), 116; X, 353.
22. See Bellinato.
23. SN 109n; X, 442.
24. Fantoli, (transl. Coyne), 121; X, 484 and 485.
25. Fantoli, (transl. Coyne), 121–122; X, 499.
26. XI, 119. (Also quoted slightly inaccurately by Koestler, 432.)
27. Fantoli, (transl. Coyne), 124–125; III Part I, 290.
28. Biagioli, 64, XI, 176.
29. XI, 241–242.
30. V, 281.
31. XII, 130.
32. Appendix I to Blackwell, *Galileo, Bellarmine and the Bible*, p. 183.
33. Fantoli, 184, XII, 172
34. Fantoli, (transl. de Santillana), 213, XII, 207.
35. Fantoli, (transl. de Santillana), 215, XIX, 320.
36. Maurice A. Finocchiaro, ed. and transl., *The Galileo Affair*, Berkeley: University of California Press, 1989, p. 146.
37. Fantoli, (transl. Coyne), 216, XIX, 321.
38. Redondi, 7.
39. Fantoli, 219–220, XIX, 321–322.
40. Fantoli, (transl. de Santillana), 223, XIX, 323.
41. Fantoli, 228, XIX, 348.
42. Fantoli, (transl. de Santillana), 225, XIX, 242.
43. Campanella, 14.
44. Campanella, 19.
45. Fantoli, (transl. de Santillana), 272–273, XII, 390–391.
46. XI, 529.
47. Camporesi, 64.
48. XII, 494.
49. Fantoli, (transl. de Santillana), 282, VI, 145–146.

50. Fantoli, (transl. Coyne), 286, XIII, 119.
51. Fantoli, (transl. Coyne), 287, XIII, 130–131.
52. Fantoli, (transl. Coyne), 294, VI, 366.
53. Fantoli, (transl. Coyne), 289, VI, 221.
54. VI, 226.
55. Drake, *Discoveries*, 278; VI, 352.
56. Fantoli, (transl. Coyne), 295, XIII, 146–147.
57. VI, 279–281.
58. Fantoli, (transl. Coyne), 319, XIII, 135.
59. Biagioli, 315n.
60. D 5 VII, 29.
61. D 69, VII, 94.
62. In *Lezioni americani*. Quoted here from Frova and Marenzana 45–46.
63. D 93, VII, 118.
64. D 103, VII, 128.
65. D 256, VII, 281.
66. D 341, VII, 368.
67. D 345, VII, 372.
68. D 345, VII, 372.
69. D 357, VII, 384.
70. D 358, VII, 385.
71. D 359, VII, 386.
72. D 367, VII, 394.
73. D 463, VII, 487–488.
74. D 464, VII, 488.
75. Fantoli, (transl. Coyne), 392, XIV, 367.
76. Fantoli, (transl. Coyne), 391, XIV, 360.
77. Biagioli, 337.
78. Fantoli, (transl. de Santillana), 395, XIV, 370.
79. Fantoli, (transl. Finocchiaro), 396, XIV, 372.
80. Fantoli, (transl. de Santillana), 460, XIV, 373.
81. Fantoli, 399 ff, XIV, 383 ff.
82. Fantoli, (transl. Finocchiaro), 400, XIV, 384.
83. Fantoli, (transl. Finocchiaro), 401, XIV, 384.
84. Fantoli, (transl. Finocchiaro), 403, XIV, 392.
85. Fantoli, (transl. de Santillana), 220, XIX, 321–322.
86. Fantoli, (transl. Coyne), 405–406, XIV, 407.
87. Fantoli, (transl. Coyne), 406, XIV, 410.
88. Fantoli, (transl. Coyne), 410, XIX, 333.
89. Fantoli, (transl. Coyne), 410, XIX, 281; 335.
90. Fantoli, (transl. Finocchiaro), 416, XV, 56.
91. Fantoli, (transl. Finocchiaro), 416, XV, 56.
92. Fantoli, (transl. Coyne), 417, XV, 62.
93. Fantoli, (transl. Coyne), 420, XV, 85.

94. Fantoli, (transl. de Santillana), 423, XIX, 339–340.
95. Von Gebler, 353.
96. Fantoli, (transl. de Santillana), 432, XIX, 344.
97. Fantoli, (transl. de Santillana), 440, XIX, 283.
98. Fantoli, (transl. de Santillana), 443, XIX, 361–362.
99. Fantoli, (transl. Coyne), 443, XIX, 362.
100. Fantoli, (transl. de Santillana), 445–446, XIX, 406–407.
101. Von Gebler, 373.
102. Bonelli and Shea, 50.
103. Camporesi, 53.
104. Camporesi, 55.
105. Remmert, 155.
106. Fantoli, (transl. Coyne), 460, XIV, 372.
107. Fantoli, (transl. Coyne), 454, XVI, 116–117.
108. TNS, xviii, VIII, 43.
109. TNS, 1, VIII, 49.
110. TNS, 5, VIII, 53.
111. TNS, 12.
112. TNS, 11 ff, VIII, 59 ff.
113. TNS, 31 ff, VIII, 77 ff.
114. TNS, 38, VIII, 83.
115. TNS, 42 ff, VIII, 87 ff.
116. TNS, 60 ff, VIII, 104 ff.
117. TNS, 107–108, VIII, 149–150.
118. TNS, 104, VIII, 147.
119. TNS, 130–131, VIII, 169.
120. TNS, 131–132, VIII, 170.
121. TNS, 153, VIII, 190.
122. TNS, 166–167, VIII, 202.
123. TNS, 169, VIII, 204.
124. Fantoli, (transl. Coyne), 512, XVII, 247.
125. Fantoli, (transl. Coyne), 370, XVII, 215.
126. Remmert, 129–130.
127. Remmert, 130.
128. Fantoli, (transl. Coyne), 490, XIX, 623.
129. Fantoli, (transl. Coyne), 490, XVIII, 378.
130. Fantoli, (transl. Coyne), 491–492, XVIII, 378–379.
131. Remmert, 25, XIX, 62
132. Isaachsen, 15n.
133. Santillana and Drake, 258.

Sources

Literature

Bassani, Riccardo and Fiora Bellini: *Caravaggio assassino*. Donzelli. Rome 1994.

Bedini, Silvio A.: "The instruments of Galileo Galilei". In McMullin, Ernan (ed.): *Galileo, Man of Science*, pp. 256–292.

Bellinati, Claudio: "Galileo Galilei e lo Studio di Padova". In *L'Osservatore Romano*, 2.–3. 9. 1991.

Biagioli, Mario: *Galileo, Courtier. The Practice of Science in the Culture of Absolutism*. University of Chicago Press. Chicago/London 1993.

Bialas, Volker: "Johannes Kepler". In Meÿenn, Karl von: *Die großen Physiker I*. pp. 157–169.

Blackwell, R.J.: *Galileo, Bellarmine and the Bible*, Notre Dame, In: University of Notre Dame Press 1991.

Bonelli, Maria Luisa Righini and William R. Shea: *Galileo's Florentine Residences*. Istituto e museo di storia della scienza. Florence [year unknown].

Campanella, Tommaso: *Apologia di Galileo*. Published and with a foreword by Luigi Firpo. Unione Tipografico-Editrice Torinese. Torino 1968.

Caporesi, Piero: *The Magic Harvest. Food, Folklore and Society*. Translated from Italian by Joan Krakover Hall. Polity Press. Cambridge 1998.

Castella, Gaston: *Papstgeschichte*. Band I–III. Updated and enlarged edition. Komet MA-Service und Verlagsgesellschaft. Frechen [year unknown].

Christianson, John Robert: *On Tycho's Island. Tycho Brahe and his Assistants 1570–1601*. Cambridge University Press. Cambridge 2000.

Cipolla, Carlo M.: *Faith, Reason and the Plague. A Tuscan Story of the Seventeenth Century*. Translated from Italian by Muriel Kittel. The Harvester Press. Brighton 1979.

Dante Alighieri: *Helvetet*. Translated into Norwegian by Magnus Ulleland. Gyldendal. Oslo 1993.

Davies, P.C.W. and Julian Brown (ed.): *Superstrings – A Theory of Everything?* Cambridge University Press. Cambridge/New York/Melbourne 1988.

Drake, Stillman and Charles T. Kowal: "Galileo's Sighting of Neptune". In *Scientific American* 243 (December 1980), pp. 52–59.

Drake, Stillman: *Dialogue Concerning the Two Chief World Systems*. Berkeley: University of California Press 1967.

Drake, Stillman: *Galileo Galilei et al., The Controversy of the Comets of* 1618, Philadelphia, PA: University of Pennsylvania Press 1960.

Drake, Stillman: *Galileo: Pioneer Scientist*. University of Toronto Press. Toronto/Buffalo/London 1990.

Fantoli, Annibale: *Galileo – For Copernicanism and for the Church*. Translated from Italian by George V. Coyne, S.J. Second revised edition. Vatican Observatory Publications. Vatican City State/Rome 1996.

Favaro, Antonio: *Galileo Galilei e Gustavo Adolfo di Svezia. Ricerche inedite*. Tipografa del seminario. Padua 1881.

Favaro, Antonio: *L'episodio di Gustavo Adolfo di Svezia nei racconti della vita di Galileo*. Officine grafiche di C. Ferrari. Venice 1906.

Feldhay, Rivka: "The use and abuse of mathematical entities: Galileo and the Jesuits revisited". In Machamer (ed.): *The Cambridge Companion . . .*, pp. 80–145.

Finocchiaro, Maurice A. (ed. and transl.): *The Galileo Affair*. Berkeley: University of California Press 1989.

Frova, Andrea and Mariapiera Marenzana: *Parola di Galileo*. Rizzioli. Milan 1998.

Galilei, Galileo: *Dialogue Concerning the Two Chief World Systems*. Translated, with revised notes, by Stillman Drake. Foreword by Albert Einstein. Second revised edition (1967). 11th impression used here: University of California Press. Berkeley, Los Angeles [year unknown].

Galilei, Galileo: *Dialogue Concerning Two New Sciences*. Translated by Henry Crew and Alfonso de Salvio. With an introduction by Antonio Favaro. Dover Publications. New York [year unknown].

Galilei, Galileo: *Il Saggiatore*. In volume II of *Opere*, published and with an introduction by Seb. Timpanaro. Rizzoli & C., Editori. Milan-Rome 1938.

Galilei, Galileo: *Le Opere di Galileo Galilei* volumes I–XX. Published by Antonio Favaro. (Re-issue of the *Edizione Nazionale*.) G. Barbèra Editore. Florence 1968.

Galilei, Galileo: *Sidereus Nuncius or The Siderial Messenger.* Translated, annotated and with an introduction and postscript by Albert van Helden. University of Chicago Press. Chicago 1989.

Garstein, Oskar: *Klosterlasse. Stormfuglen som ville erobre Norden for katolisismen.* Aschehoug/Thorleif Dahls kulturbibliotek. Oslo 1998.

Gebler, Karl von: *Galileo Galilei. Leben und Werk.* (Published by G. Peers from the original publication of 1875). Emil Vollmer Verlag. Essen [year unknown].

Ginzburg, Carlo: *The Cheese and the Worms. The Cosmos of a Seventeenth-century Miller.* Translated from Italian and with a preface by John and Anne Tedeschi. Penguin Books. London 1992.

Gregori, Mina: *Caravaggio. Come nascono i capolavori.* Electa. Milan 1991.

Grell, Ole Peter and Bob Schribner (ed.): *Tolerance and Intolerance in the European Reformation.* Cambridge University Press. Cambridge 1996.

Hamel, Jürgen: "Nicolaus Copernicus". In Meÿenn, Karl von: *Die großen Physiker I*, pp. 131–145.

Heer, Friederich: *The Holy Roman Empire.* Translated from German by Janet Sondheimer. Phoenix. London 1995.

Hellesnes, Jon: *René Descartes.* Gyldendal (Ariadne Series). Oslo 1999.

Hibbert, Christopher: *The Rise and Fall of the House of Medici.* Penguin Books. London 1985.

Isaachsen, D. (ed. Johan Holtsmark): *Lærebok i fysikk for realgymnaset I.* 16th edition, second impression. Aschehoug. Oslo 1965.

Jardine, Lisa: *Ingenious Pursuits.* Little, Brown and Company. London 1999.

Koestler, Arthur: *The Sleepwalkers. A History of Man's Changing Vision of the Universe.* (1959) Penguin Books. London 1977.

Krafft, Fritz: "Aristoteles". In Maÿenn, Karl von: *Die großen Physiker I*, pp. 78–101.

Lagerkvist, Lars O.: *Sverige och dess regenter under 1000 år.* Albert Bonniers Förlag AB. Stockholm 1982.

Machamer, Peter (ed.): *The Cambridge Companion to Galileo.* Cambridge University Press. Cambridge/New York/Melbourne 1998.

Machamer, Peter: "Galileo's Machines, his mathematics, and his experiments". In Machamer, Peter (ed.): *The Cambridge Companion...* pp. 53–79.

McMullin, Ernan (ed.): *Galileo, Man of Science.* Basic Books, Inc. New York/London 1967.

Mereu, Italo: *Storia dell'Intolleranza in Europa.* Tascabili Bompiani. Milan 2000 (sixth impression).

Meÿenn, Karl von (publisher): *Die großen Physiker. I–II.* Verlag C.H. Beck. Munich 1997.
Michelsen, Karin (ed.): *Cappelens musikkleksikon.* J.W. Cappelens forlag. Oslo 1979.
Montanelli, Indro and Roberto Gervaso: *L'Italia del Seicento.* Rizzoli Editore. Milan 1969, 1998.
Panofsky, Erwin: *Galileo as a Critic of the Arts.* Martinus Nijhoff. The Hague 1954.
Pustka, Josef: *Basilica di Santa Maria Maggiore.* D.EDI.T s.r.l. Rome 1997.
Ravaglioli, Armando: *Breve storia di Roma dalle origini ai giorni nostri.* Tascabili economici Newton (second edition). Rome 1995.
Redondi, Pietro: "From Galileo to Augustine". In Machamer, Peter (ed.): *The Cambridge Companion . . .*, pp. 175–210.
Redondi, Pietro: *Galileo, Heretic.* Translated from Italian by Raymond Rosenthal. Princeton University Press. Princeton, New Jersey 1987.
Remmert, Volker R.: *Ariadnefäden im Wissenschaftslabyrinth.* Studien zu Galilei: Historiographie – Mathematik – Wirkung. Peter Lang. Bern 1998.
Reston, James: *Galileo: A Life.* Harper Collins Publishers. New York 1994.
Ricci, Saverio: *Giordano Bruno nell'Europa del Cinquecento.* Salerno. Rome 2000.
Rystad, Göran: *Religionskriger og enevelde.* Translated from Swedish [to Norwegian] by Egil A. Kristoffersen. (Volume 11 of *Cappelens verdenshistorie*, ed. Erling Bjøl.) J.W. Cappelens Forlag A/S. Oslo 1985.
Santillana, Giorgio de: *The Crime of Galileo*, Chicago: University of Chicago Press 1955.
Santillana, Giorgio de and Stillman Drake: "Arthur Koestler and his Sleepwalkers". In: *Isis* 50 (1959) pp. 255–260.
Sharratt, Michael: *Galileo, Decisive Innovator.* Cambridge University Press. Cambridge/New York/Melbourne 1994.
Singh, Simon: *Fermat's Last Theorem.* Fourth Estate. London 1997.
Sobel, Dava: *Galileo's Daughter.* Fourth Estate. London 1999.
Sobel, Dava: *Longitude.* Fourth Estate. London 1996.
Teres, Gustav: *The Bible and Astronomy.* Springer Orviosi Kiadó Kft. Budapest 2000.
Turner, Jane (ed.): *A Dictionary of Art.* Macmillan. London 1996.
Vannucci, Marcello: *The History of Florence.* Translated from Italian by Charles Lambert. Newton Compton Editori. Rome 1988.
Wallace, William A.: "Galileo's Pisan studies in science and philosophy". In Machamer (ed.): *The Cambridge Companion . . .*, pp. 27–52.

Wallace, William A.: *Galileo's Logic of Discovery and Proof.* Kluwer Academic Publishers. Dordrecht/Boston/London 1992.
White, Michael: *Leonardo. The First Scientist.* Little, Brown and Company. London 2000.
Wootton, David: *Paolo Sarpi. Between Renaissance and Enlightenment.* Cambridge University Press. Cambridge/New York/Melbourne 1983.
Yates, Frances A.: *Giordano Bruno and the Hermetic Tradition* (1964). 10th impression used here: University of Chicago Press. Chicago 1999.

Internet

The Galileo Project. Rice University 1995 –
Lead by Albert van Helden and Elizabeth Burr
http://es.rice.edu/ES/humsoc/Galileo
The Catholic Encyclopedia
http://www.newadvent.ord/cathen
Università di Padova
http://www.unipd.it

Index of Names

Printing: Krips bv, Meppel
Binding: Litges & Dopf, Heppenheim